I0064621

INTRODUCTION TO ELECTROWEAK UNIFICATION

Standard Model from Tree Unitarity

INTRODUCTION TO ELECTROWEAK UNIFICATION

Standard Model from Tree Unitarity

J. Horejsi
Charles University

World Scientific
Singapore • New Jersey • London • Hong Kong

Published by

World Scientific Publishing Co. Pte. Ltd.
5 Toh Tuck Link, Singapore 596224
USA office: 27 Warren Street, Suite 401-402, Hackensack, NJ 07601
UK office: 57 Shelton Street, Covent Garden, London WC2H 9HE

British Library Cataloguing-in-Publication Data
A catalogue record for this book is available from the British Library.

INTRODUCTION TO ELECTROWEAK UNIFICATION — STANDARD MODEL FROM TREE UNITARITY

Copyright © 1994 by World Scientific Publishing Co. Pte. Ltd.

All rights reserved. This book, or parts thereof, may not be reproduced in any form or by any means, electronic or mechanical, including photocopying, recording or any information storage and retrieval system now known or to be invented, without written permission from the publisher.

For photocopying of material in this volume, please pay a copying fee through the Copyright Clearance Center, Inc., 222 Rosewood Drive, Danvers, MA 01923, USA. In this case permission to photocopy is not required from the publisher.

ISBN-13 978-981-02-1857-7
ISBN-10 981-02-1857-5

CONTENTS

v

Appendices

PREFACE

In this work I have presented a non-traditional introduction to the theory of unification of weak and electromagnetic interactions. In contrast to the usual textbook treatments I describe here in detail a derivation of the standard model of electroweak interactions based on a straightforward application of the requirement of perturbative renormalizability. A necessary condition for perturbative renormalizability is supposed to be the corresponding ("unitary") behaviour of the *tree-level* Feynman diagrams in high-energy limit (a technical term "tree unitarity" is commonly used for such a condition in current literature).

It is well known that the contemporary standard model of electroweak interactions was formulated in the 1960's by S. Glashow, S. Weinberg and A. Salam who employed the principles of non-abelian gauge invariance and Higgs mechanism. The road to the standard model described in the present text was discovered somewhat later (during the first half of the 1970's) and reported in the papers [11 14] and its most remarkable feature is that it demonstrates the *necessity* of the original principles if perturbative renormalizability of the S-matrix is to be achieved.

It should be emphasized, however, that the requirement of perturbative renormalizability in fact does not represent an "absolute dogma" for constructing a realistic theory; an experimental verification of predictions of a renormalizable theory only means that conceivable interactions of a non-renormalizable type may play a role on a distant, so far inaccessible energy scale (for a discussion of the problem of renormalizability from a modern point of view see e.g. Ref. [72]). Actually, nowadays there seems to be a widespread belief that the Glashow-Weinberg-Salam (GWS) standard model is merely an effective theory (which is phenomenologically successful in an accessible energy region). In other words, it is most probably just a "low-energy approximation" of a deeper theory. There are several alternatives (see e.g. [73 75]), yet the existing experimental data do not indicate any clear direction.

Anyway, it is clear that the requirement of perturbative renormalizability (which may now be regarded as a constraint of a rather technical nature) played the role of an extremely useful heuristic principle in the theory of weak and electromagnetic interactions, since the GWS theory led to many highly non-trivial predictions, a significant part of which have already been confirmed by experiments. Thus, one may say that regardless of a future development of our ideas (in particular concerning an essence of the Higgs mechanism) the GWS standard model will remain a relevant part of particle physics, not only as a phenomenologically successful effective theory valid in certain energy region, but also as a construction which is remarkable from a purely theoretical, methodical point of view.

The present text originates from a series of lectures for graduate students specialized in theoretical physics and particle physics which I delivered in the period 1986 - 1992 at the Faculty of Mathematics and Physics of the Charles University in Prague (these lectures form a part of a one-semester course). The main reason for transforming my handwritten notes into this text was the fact that the diagrammatic derivation of the GWS standard model from the requirement of tree unitarity (i.e. of a good high-energy behaviour of tree-level scattering amplitudes) is not covered by the existing textbooks and monographs in sufficient detail. At the same time this approach, which is consequently deductive and systematic, is also quite straightforward and instructive and thus it seems to be attractive even from the point of view of pedagogical clarity. The conventional formulation of the standard model as a non-abelian gauge theory with the Higgs mechanism is not given here, as it can be found in many textbooks such as e.g. [17], [21], [25], [36], [56], [76], [77]. The text is divided into five chapters; the first four of them have in a sense preparatory character as there are discussed mostly the difficulties of provisional (non-renormalizable) models of weak and electromagnetic interactions which ultimately lead to the need for unification of both forces. The core of the whole work is Chapter 5 where the diagrammatic construction of electroweak unification (i.e. the above mentioned "non-standard derivation of the standard model") is described in detail. Our exposition in that chapter is close in spirit e.g. to the article of S. Joglekar [14] and also to the lecture notes of C. H. Llewellyn Smith [18] and R. Kleiss [39] (Ref. [18] has been particularly stimulating); however, it is essentially independent of these treatments and is also more detailed in some respects. The main text is supplemented by a series of technical appendices which should further minimize the dependence of the whole work on external sources. The text should thus be digestible even for an uninitiated reader; a necessary prerequisite is only an elementary knowledge of quantum field theory on the level of Feynman diagrams and also some familiarity with the basic concepts of particle physics, including in particular the Fermi–Feynman–Gell–Mann $V - A$ model of weak interactions. I believe that the present work may also be useful for a more experienced reader familiar with the conventional formulation of the standard model; it turns out that details of the "diagrammatic" derivation based on tree unitarity are relatively little known in comparison with the traditional approach. Section 5.6 devoted to the effects of the Adler-Bell-Jackiw anomaly goes slightly beyond the basic framework of the main text (a rather detailed discussion contained there reflects to some extent the author's own predilection in the subject of anomaly). However, a detailed knowledge of the material of Section 5.6 is not necessary for understanding of the bulk of Chapter 5; what really matters for the first reading is just the simple formula (5.119) which is also needed later in Section 5.7. Each chapter is also supplemented by exercises and problems.

Finally, a remark on the cited literature is in order. I did not attempt to present a full list of literature concerning the standard model in the present context; only the works necessary for the purpose of reference are included here. In this connection, the reader may find particularly useful the book [77] which contains an extensive list of relevant literature.

At this place I would like to express my thanks to Dr. M. Jirásek for checking some of the exercises. My thanks are also due to Dr. P. Kolář for a valuable comment on the proof of the main statement of Appendix I. I am also grateful to students and other participants of lectures and seminars at Prague University and the Institute of Physics of the Czech Academy of Sciences for all the discussions and comments which helped to improve the present text. I wish to thank Mrs. L. Hiršlová for excellent typing of the manuscript. I am greatly indebted to Mrs. Sue Chinnick for careful proofreading of the text. Last but not least, I would like to thank Mr. M. Stöhr for invaluable technical assistance in preparing the final form of the manuscript.

Prague, April 1994 J. Hořejší

Conventions and Notation

Unless stated otherwise, we always use the natural system of units in which $\hbar = c = 1$.

Most of the other conventions correspond to the textbook of Bjorken and Drell [16]. The indices of any Lorentz four-vector or tensor take on values 0, 1, 2, 3. The metric is defined by

$$g_{\mu\nu} = \text{diag}(+1, -1, -1, -1),$$

so that e.g. the scalar product $k.p$ is

$$k.p = k_0 p_0 - \vec{k}.\vec{p}$$

Dirac matrices γ^μ, $\mu = 0, 1, 2, 3$ are defined by means of the standard representation [16]. We also employ the usual symbol $\not{p} = p_\mu \gamma^\mu$ for an arbitrary four-vector p. We should particularly stress the definition of the γ_5 matrix:

$$\gamma_5 = i\gamma^0 \gamma^1 \gamma^2 \gamma^3$$

Further, the fully antisymmetric Levi-Civita tensor $\varepsilon_{\mu\nu\rho\sigma}$ is fixed by the convention

$$\varepsilon_{0123} = +1$$

(Let us remark that this convention differs in sign e.g. from that used by Itzykson and Zuber [21].)

Conventions for Dirac spinors are described in Appendix B. Let us emphasize that the normalization employed here differs from [16] (it coincides e.g. with [20]).

Finally, the Lorentz-invariant transition (scattering) amplitude $\mathcal{M}_{fi}$ (for brevity usually denoted simply as $\mathcal{M}$) has an opposite sign with respect to Bjorken and Drell [16] (the convention adopted here corresponds e.g. to [20]).

Frequently Used Symbols

$*$	complex conjugation (c.c.)
$\dagger$	hermitian conjugation (h.c.)
$\bar{\psi}$	Dirac conjugation ($\bar{\psi} = \psi^\dagger \gamma_0$)
A_μ	four-potential of electromagnetic field
$F_{\mu\nu}$	tensor of electromagnetic field
W_μ^-	field of charged vector bosons involving annihilation operator of the particle W^-
W_μ^+	field of charged vector bosons involving annihilation operator of the particle W^+ (it holds $W_\mu^+ = (W_\mu^-)^\dagger$)
$W_L^\pm$, Z_L	longitudinally polarized vector bosons $W^\pm$, Z
$W_T^\pm$, Z_T	transversely polarized vector bosons
η	neutral scalar (Higgs) boson or the corresponding field
$E^\pm$, E^0	heavy leptons of electron type, or the corresponding fields
f	arbitrary standard fermion (lepton or quark, or the corresponding field); exceptionally also a coupling constant
f_L, f_R	left-handed or right-handed component of a fermion f (exceptionally also coupling constants for heavy lepton interactions)
l	charged lepton ($l = e, \mu, \tau$), or the corresponding field
ν_l, $\bar{\nu}_l$	neutrino (antineutrino) corresponding to the lepton l, or the corresponding field
e^-, e^+	electron, positron
ν, $\bar{\nu}$	neutrino (antineutrino) of the electron type, or the corresponding field
u, c, t	quarks with charge 2/3, or corresponding fields
d, s, b	quarks with charge $-1/3$, or corresponding fields
γ	photon
G_F	Fermi coupling constant
e	electromagnetic coupling constant (positron charge)
α	fine-structure-constant ($\alpha = e^2/4\pi \doteq 1/137$)
g	coupling constant for interactions of weak charged currents and $W^\pm$
ϑ_C	Cabibbo angle
V_{CKM}	Cabibbo-Kobayashi-Maskawa matrix

ϑ_W	parameter of interactions of weak neutral currents ("weak mixing angle", "Weinberg angle")
$\varepsilon_L^{(f)}$, $\varepsilon_R^{(f)}$	parameters of interactions of left-handed, or right-handed components of neutral current corresponding to a fermion f
v_f, a_f	parameters of interactions of vector or axial-vector components of neutral current corresponding to a fermion f
Q_f	charge of a fermion f in units of e
$u(p)$, $v(p)$	Dirac spinor for a fermion or antifermion, with four-momentum p
$\varepsilon(p)$, $\varepsilon^\mu(p)$	four-vector of an (arbitrary) polarization of vector boson with four-momentum p
$\varepsilon_L(p)$, $\varepsilon_L^\mu(p)$	four-vector of longitudinal polarization
$\varepsilon_T(p)$, $\varepsilon_T^\mu(p)$	four-vector of transverse polarization
$\mathcal{V}_{\lambda\mu\nu}(k,p,q)$, $V_{\lambda\mu\nu}(k,p,q)$	interaction vertex WWZ or $WW\gamma$
s, t, u	Mandelstam kinematical invariants
$E_{c.m.}$	centre-of-mass energy ($E_{c.m.} = s^{1/2}$)
E	typical energy of a considered process (e.g. $E_{c.m.}$)
Ω	solid angle

Chapter 1

INTRODUCTION

One of the cornerstones of particle physics in the early 1960's was a phenomeno-logically successful theory of weak interactions based on Fermi's original idea [1] of a direct interaction of four spin-$\frac{1}{2}$ fields. A decisive role in formulating this theory can be attributed to Feynman and Gell-Mann [2]; an important improvement of the Feynman – Gell-Mann theory is due to Cabibbo [3]. The corresponding interaction Lagrangian may be written as

$$\mathcal{L}_{int}^{(w)} = -\frac{G_F}{\sqrt{2}} J^\rho J_\rho^\dagger \tag{1.1}$$

where G_F is the Fermi coupling constant determined from the measured lifetime of muon, $G_F \doteq 1.166 \times 10^{-5} \mathrm{GeV}^{-2}$. The current J^ρ has lepton and hadron parts,

$$J^\rho = J_{(lepton)}^\rho + J_{(hadron)}^\rho \tag{1.2}$$

where (taking into account only the leptons e, ν_e, μ, ν_μ)

$$J_{(lepton)}^\rho = \bar{\nu}_e \gamma^\rho (1 - \gamma_5) e + \bar{\nu}_\mu \gamma^\rho (1 - \gamma_5) \mu \tag{1.3}$$

and the hadron part can be expressed in modern language by means of quark fields (if we consider also the c-quark beside the u, d, s)

$$\begin{aligned} J_{(hadron)}^\rho &= \bar{u} \gamma^\rho (1 - \gamma_5)(d \cos \vartheta_C + s \sin \vartheta_C) \\ &+ \bar{c} \gamma^\rho (1 - \gamma_5)(-d \sin \vartheta_C + s \cos \vartheta_C) \end{aligned} \tag{1.4}$$

where ϑ_C is the Cabibbo angle ($\vartheta_C \approx 13°$). (However, one should keep in mind that the relevance of the c-quark was confirmed only in mid 1970's; the original Cabibbo current was given, roughly speaking, only by the first term in (1.4).)

In the commonly used terminology, the Lagrangian (1.1) corresponds to an in-teraction of two "charged" currents; the technical term "charged current" simply means that in the expressions (1.3) or (1.4) occur pairs of fields with different charge ($(\nu_e, e), (u, d)$ etc.). (For example, the electromagnetic current is then "neutral", in the sense of this terminology.) From the point of view of space-time symmetries, the current (1.2) is of the type $V - A$, i.e. it is a Lorentz vector minus an axial vector (pseudovector). In other words, only left-handed parts of fermion fields (e.g. $e_L = \frac{1}{2}(1 - \gamma_5)e$ etc.) participate in weak interactions (this in fact was the original hypothesis proposed by Feynman and Gell-Mann [2]). This corresponds to a maximum violation of parity in the Lagrangian (1.1): The parity-violating interaction (term VA and AV) and the parity-conserving interaction (terms VV and AA) have an equal strength and this in turn leads to maximum parity-violating effects in observable quantities.

The theory described by the relations (1.1) - (1.4) is usually called the phenomenological (or effective) $V - A$ theory of weak interactions. The adjectives "phenomenological" or "effective" reflect the fact that this theory described well most of the relevant experimental data known in the 1960's but the calculations of decay rates and cross sections of physical processes were only practicable on the level of tree Feynman diagrams (i.e. those not involving closed loops of internal lines) since the higher-order contributions in the perturbation expansion were not renormalizable by means of the standard methods (in contrast with e.g. quantum electrodynamics). Moreover, it also soon became clear that the approximation of tree diagrams can reasonably describe weak scattering processes only for sufficiently low energies of the interacting particles; a typical order-of-magnitude estimate amounts to

$$E_{c.m.} = s^{\frac{1}{2}} \ll G_F^{-\frac{1}{2}} \doteq 300\text{GeV} \tag{1.5}$$

where $E_{c.m.}$ is the corresponding collision energy in the centre-of-mass (c.m.) system.

The above-mentioned difficulties of the four-fermion weak interaction theory (1.1) within the perturbative framework (i.e. the non-renormalizability of the closed-loop diagrams and the inapplicability of the tree approximation at high energies) had a purely theoretical character in the 1960's. However, these technical flaws indicated that such a model, though phenomenologically successful practically until the early 1970's, did not provide a full theory of weak interactions and could only represent a certain approximation to a fundamental theory in the low-energy limit.

The road to a more satisfactory (i.e. renormalizable) theory of weak interactions is remarkable in itself both historically and methodically, as it was based substantially on a development of new ideas and techniques in field theory. From the physical point of view, it is interesting mainly because it has finally led to a model which in a sense unifies weak and electromagnetic interactions and provides some highly non-trivial theoretical predictions, a part of which has already been verified experimentally. The history of the discovery of the renormalizable unified theory of weak and electromagnetic interactions has been described brilliantly by S. Weinberg, A. Salam and S. Glashow in their Nobel lectures [4].

Glashow-Weinberg-Salam (GWS) theory [5, 6, 7] is based on the principles of non-abelian gauge invariance (i.e. the Yang-Mills field) [8] and Higgs mechanism [9]. The renormalizability of non-abelian gauge theories with the Higgs mechanism was proved by 't Hooft in 1971 [10] and experimental evidence supporting the validity of the GWS model has been accumulating continually since the early 1970's (when the weak neutral currents were discovered). In view of its phenomenological successes the GWS theory is now usually called the standard model of electroweak interactions (this term became widely recognized during the 1980's). A major triumph of the standard GWS model then has been the discovery of the intermediate vector bosons W and Z (in 1983) possessing the properties predicted by the theory. In a sense, a "new era" in the physics of electroweak interactions started in 1989 in connection with the launching of experiments on the electron-positron colliders LEP at CERN (Geneva) and SLC (Stanford, USA). These new precision measurements now make it possible to verify even the theoretical predictions of higher-order perturbative effects (denoted generally as "radiative corrections"). It is expected that experiments on these colliders and on the others now under consideration will make it possible ultimately to test the correctness of basic principles of the standard model, i.e. the non-abelian gauge symmetry and the Higgs mechanism, by the end of the 1990's.

In subsequent chapters we describe a road leading from the Feynman – Gell-Mann model of the four-fermion $V - A$ interaction (1.1) to the GWS standard model. In contrast to most of the existing literature, in this text we present a derivation of the standard model based on the requirement of "tree unitarity" (i.e. an "asymptotic softness" of scattering amplitudes corresponding to *tree-level* Feynman diagrams in high-energy limit); such a requirement is in fact a necessary condition of the perturbative renormalizability at higher orders. This alternative approach is rather straightforward and instructive, and what is most important, it demonstrates the necessity of non-abelian gauge fields and also the inevitability of a scalar Higgs boson in the renormalizable theory of weak interactions. Such a derivation of the standard model appeared in the literature somewhat later than the original GWS construction (see [11 14]). In the present work we give a detailed treatment of this diagrammatic approach in a form which should be digestible even for an uninitiated reader unacquainted with the traditional "textbook" formulation of the standard model of electroweak interactions.

Chapter 2

DIFFICULTIES OF FERMI THEORY

2.1 Perturbative Non-Renormalizability

Some technical background for this chapter may be found in the Appendices A - G.

If one considers a general Feynman diagram in a Fermi-type theory of weak interactions, i.e. in a theory of direct four-fermion interaction (exemplified by (1.1)), then the corresponding superficial degree of divergence (i.e. the ultraviolet divergence "index") is given by the formula (G.8) of Appendix G where the relevant index of the four-fermion interaction vertex is $\omega_v = 6$ (this is obtained by setting $n_F = 4$, $n_B = 0$, $n_D = 0$ in the formula (G.9)). This indicates that a direct (contact) four-fermion interaction leads to non-renormalizable perturbation expansion, since by iterating the four-fermion vertex in Feynman diagrams to a sufficiently high order one may expect ultraviolet divergent graphs to occur for an arbitrary configuration of external lines, i.e. one might encounter an infinite number of divergence types which in turn would require an infinite number of renormalization counterterms. A more detailed analysis indeed confirms such an expectation (see e.g. Ref. [4]). It is also obvious that in the considered case the value of the index $\omega_v = 6$ is closely related to the fact that the dimension of the Fermi coupling constant G_F is M^{-2}, in units of an arbitrary mass M (cf. Appendix G, the discussion around the relation (G.11)).

2.2 Tree-Level Violation of Unitarity

In view of the inapplicability of standard methods of quantum field theory in higher orders of perturbation expansion, we may restrict ourselves to the lowest order only - i.e. to the approximation of tree diagrams. We shall consider the purely leptonic sector of the theory described by the interaction Lagrangian (1.1), i.e.

$$\mathcal{L}^{(w)}_{(lepton)} = -\frac{G_F}{\sqrt{2}} J^\rho_{(lepton)} J^\dagger_{\rho(lepton)} \tag{2.1}$$

4

where the current $J^\rho_{(lepton)}$ is defined by the expression (1.3). Let us now investigate in more detail the elastic scattering processes $\nu_e e \to \nu_e e$ and $\bar\nu_e e \to \bar\nu_e e$ in the high-energy limit, i.e. for $E_{c.m.} \gg m_e$ (in what follows the index e is usually omitted for brevity). It can be expected (and it is indeed confirmed by an explicit calculation - see Appendix D) that in such a limit one may neglect m_e. Asymptotic behaviour of the corresponding amplitudes and cross sections may be then estimated on the basis of simple dimensional considerations: In the system of units we are using ($\hbar = c = 1$) a cross section has dimension $[\sigma] = M^{-2}$ (i.e. (energy)$^{-2}$) and in the lowest order, i.e. in the 1st order of perturbation expansion with respect to the interaction (2.1), it must be proportional to G_F^2. Taking into account that G_F has the dimension of (energy)$^{-2}$ and neglecting the effects of masses of the interacting particles, the integral cross section can then only depend on the kinematical invariant s (see Appendix A, definition (A.3)). It is clear that the only quantity with the dimension of a cross section and proportional to G_F^2 is (up to a dimensionless constant) $G_F^2 s$. Thus one may expect that in the limit $s \to \infty$ the cross section of the process $\nu e \to \nu e$ behaves like

$$\sigma(\nu e \to \nu e) \simeq const. \times G_F^2 s \qquad (2.2)$$

and similarly for $\bar\nu e \to \bar\nu e$. The estimate (2.2) is confirmed by an explicit calculation performed in Appendix D which gives the results (see (D.13), (D.14))

$$\sigma(\nu e \to \nu e) = \frac{1}{\pi} G_F^2 s \qquad (2.3)$$

$$\sigma(\bar\nu e \to \bar\nu e) = \frac{1}{3\pi} G_F^2 s \qquad (2.4)$$

if one neglects m (we always tacitly assume such an approximation unless stated otherwise). Analogous dimensional considerations lead to the conclusion that the corresponding scattering amplitude $\mathcal{M}_{fi}$ (which is dimensionless for binary processes see Appendix C, formula (C.3)) behaves (for a fixed scattering angle) like $G_F s$ in the tree approximation.

We thus see that in a Fermi-type theory of weak interactions the scattering amplitudes and cross sections calculated from tree diagrams rise linearly with s (i.e. quadratically with the centre-of-mass energy).

However, for sufficiently high energies, such behaviour leads to an apparent conflict of perturbative (tree-level) approximation with a general property of the exact S-matrix, namely with unitarity. The explanation of such a remarkable statement is quite simple if we use a partial-wave expansion of the relevant amplitude or cross section (see Appendix E). Indeed, if a (tree-level) scattering amplitude $\mathcal{M}(s,\Omega)$ depends linearly on s (like $G_F s$) then an analogous unbounded growth for $s \to \infty$ may be expected for the corresponding partial-wave amplitudes as well (cf. (E.7)). Thus, for sufficiently large values of s (of an order $s \geq G_F^{-1}$) the tree approximation will violate the unitarity condition (cf. (E.12))

$$|\mathcal{M}^{(j)}(s)| \leq 1 \qquad (2.5)$$

Let us now illustrate this simple qualitative consideration by a concrete example of the process $\nu e \to \nu e$. If we neglect the electron mass, a corresponding scattering amplitude is non-zero only for the combination of helicities

$$h_1 = h_2 = h_1' = h_2' = -\frac{1}{2} \tag{2.6}$$

(this is a consequence of the $V - A$ structure of charged currents in the interaction (2.1)). From the result of the calculation performed in Appendix D (see the formula (D.5)) then it immediately follows that for the helicities (2.6) one has

$$|\mathcal{M}_{h_1' h_2' h_1 h_2}(s, \Omega)| = 4\sqrt{2}\, G_F s \tag{2.7}$$

Comparing (2.7) with the general formula (E.6) and taking into account the relation (F.4) from Appendix F we see that for such a combination of helicities the Jacob-Wick expansion is in fact an expansion in Legendre polynomials (as $\lambda = h_1 - h_2 = 0$, $\lambda' = h_1' - h_2' = 0$) and the independence of (2.7) on the scattering angle implies that only the partial wave with $j = 0$ contributes. For the amplitude of this partial wave we then get immediately

$$|\mathcal{M}^{(0)}(s)| = \frac{1}{2\sqrt{2}\,\pi}\, G_F s \tag{2.8}$$

and for the cross section (corresponding to the combination of helicities (2.6)) we have

$$\sigma = \sigma^{(0)} = \frac{2}{\pi} G_F^2 s \tag{2.9}$$

The unitarity condition (2.5) (or (E.19)) then gives the bound $s \leq 2\pi\sqrt{2}\, G_F^{-1}$, i.e.

$$E_{c.m.} = \sqrt{s} \leq \left(\frac{2\pi\sqrt{2}}{G_F}\right)^{\frac{1}{2}} \tag{2.10}$$

The critical value $\sqrt{s_0}$, for which the unitarity condition is saturated (i.e. such that in (2.5) or (E.19) the equality holds) is usually called "unitarity bound" (see e.g. [15]) since for $E_{c.m.} > \sqrt{s_0}$ the tree approximation (2.8) (or (2.9)) violates a necessary condition of unitarity and thus obviously ceases to be a good approximation. In the considered particular case the corresponding value (see (2.10)) is $\sqrt{s_0} = (2\pi\sqrt{2})^{\frac{1}{2}} G_F^{-\frac{1}{2}} \approx 870$ GeV. Of course, the value of a unitarity bound is process-dependent (see problems 2.2 and 2.3 at the end of this chapter).

It is in order to emphasize here that the violation of unitarity discussed in this chapter refers to the lowest perturbative order; the exact S-matrix (if we were able to calculate it) should of course be unitary as the hamiltonian is hermitian.

It is easy to understand that the S-matrix calculated to a finite order of perturbation expansion is not unitary, if one realizes that the unitarity condition $SS^+ = S^+S = 1$ is nonlinear and thus it connects contributions of different perturbative order (see e.g. [16], Chapter 8). Thus, in the considered case of the four-fermion interaction, the tree-level S-matrix is in fact not unitary for any value of the energy of interacting particles just because we are neglecting higher-order contributions. For sufficiently low energies ($G_F s \ll 1$) the tree-level S-matrix differs little from a unitary matrix; a possible deviation from unitarity is of the order $O(G_F^2 s^2)$ and it is not possible to draw any conclusion from the simple criterion expressed by the inequality (2.5) (this inequality is only a *necessary* condition of the unitarity). However, a unitarity violation is manifest for sufficiently high energies (such that $G_F s \geq 1$) when the condition (2.5) is no longer satisfied. One may then also expect that the deviation from unitary behaviour in the tree approximation is substantial, of the order $O(1)$.

2.3 High-Energy Behaviour and Renormalizability

It is important to realize that the inequality (2.5) is in general violated (for sufficiently high energies) even for tree-level scattering amplitudes of spinor electrodynamics, although in some particular cases the condition (2.5) may accidentally be satisfied for an arbitrary energy (see the problems 2.4 and 2.5 at the end of this chapter). However, in contrast to the four-fermion weak interaction model, the corresponding amplitudes of partial waves in spinor QED grow at most *logarithmically* with energy; this turns out to be a behaviour typical for perturbatively *renormalizable* theories (see e.g. [12], [17], [18]). (Let us also stress, in connection with the problem 2.5, that spinor QED is renormalizable even in the case that "photon" has a non-zero mass - see e.g. [17], [21]).

As we have seen, applications of the perturbation expansion in a theory of the Fermi type face two problems:

1. The perturbation series is not renormalizable by means of standard methods.

2. Scattering amplitudes corresponding to tree diagrams grow with energy as $E_{c.m.}^2$ and for $E_{c.m.} \geq G_F^{-\frac{1}{2}}$ (i.e. for high, but still "terrestrial" energies) the tree approximation is manifestly inapplicable.

We have already mentioned that these two problems are in fact closely related to each other. More precisely, *a power-like growth of tree-level amplitudes with respect to energy seems to imply non-renormalizability at higher orders of perturbation expansion*. This remarkable connection of two different aspects of perturbation expansion will be a subject of more detailed considerations in subsequent chapters and at the same time it will serve as an important heuristic principle leading eventually to a realistic theory of weak interactions.

Of course, it is highly desirable to have a renormalizable model of weak interactions, i.e. to have a theory comparable with e.g. spinor QED. From what we have already said it follows immediately that for this purpose one has to look for an adequate model of quantum field theory, in which tree-level scattering amplitudes do not exhibit a power-like growth with energy. Tree approximation will then also be applicable in a much wider range of energies than in the case of a Fermi-type theory.

The model with charged intermediate vector boson, described in the next chapter, solves the problem of high-energy behaviour of tree-level scattering amplitudes only for some processes (e.g. for neutrino-electron scattering in particular); nevertheless, it is an important first step towards a renormalizable theory of weak interactions.

Problems

2.1. Calculate cross sections of scattering processes $\nu_e e \to \nu_e e$ and $\bar\nu_e e \to \bar\nu_e e$ (in the lowest order of perturbation expansion) in the high-energy limit (i.e. neglecting m_e) on the assumption that weak lepton current has the form $vV - aA$ (i.e. it involves the combination of Dirac matrices $\gamma_\rho(v - a\gamma_5)$), where a, v are real constants. Show that for arbitrary values of a, v one has, in the considered approximation

$$2 \le \frac{\sigma^{(\nu e)}(s)}{\sigma^{(\bar\nu e)}(s)} \le 3$$

2.2. Calculate "unitarity bounds" for processes $\bar\nu_e e \to \bar\nu_e e$ and $e^- e^+ \to \nu_e \bar\nu_e$ within the framework of the Feynman – Gell-Mann (FGM) model of weak interactions with $V - A$ currents.

2.3. For which lepton processes (admissible in the lowest perturbative order in FGM model) has the unitarity bound the maximum and minimum value respectively?

2.4. Consider the process $e^- e^+ \to \mu^- \mu^+$ in the framework of spinor QED in the high-energy limit, i.e. for $s \gg m_\mu^2$. Which partial waves contribute to the corresponding tree-level amplitude in the Jacob-Wick expansion? What restrictions are imposed by unitarity in this case?

2.5. Discuss the partial-wave expansion of the tree-level scattering amplitude for $e\mu \to e\mu$ in high-energy limit. Assume that the photon has a non-zero mass. What role does the photon mass play in the calculation of partial-wave amplitudes?

Chapter 3

INTERMEDIATE VECTOR BOSON

3.1 Hypothesis of Charged Massive IVB

A necessary technical background for this chapter may be found in Appendix H.

One of the important results of the preceding chapter is an observation that difficulties of the weak interaction theory of the Fermi type are intimately related to the contact character (i.e. zero range) of the four-fermion interaction described by the Lagrangian (1.1): It is precisely the assumption of direct interaction of four fermion fields which causes the corresponding coupling constant (i.e. the G_F) to have the dimension of a negative power of mass.

Therefore it is natural to consider instead of (1.1) an interaction described by an "exchange" of another particle (which must then necessarily be a boson) in analogy with e.g. photon exchange in QED. (Such an idea was probably formulated for the first time by O. Klein in 1938.) In its simplest realization it means formally the passage from (1.1) to the interaction Lagrangian which may be written as

$$\mathcal{L}_{int}^{(w)} = \frac{g}{2\sqrt{2}}(J^\rho W_\rho^+ + J^{\dagger\rho} W_\rho^-) \qquad (3.1)$$

Here J^ρ is the weak current defined by relations (1.2) - (1.4) (we shall consider only its lepton part in what follows) and $W_\rho^\pm$ is the vector field corresponding to a "mediating" particle (with spin 1) which is therefore usually called intermediate vector boson (IVB). Unlike the photon (which is actually an IVB of electromagnetic interaction), the IVB of weak interactions carries an electric charge (± 1 in units of positron charge); this, of course, is due to the fact that the weak current in (3.1) is "charged" in the sense defined in Chapter 1. In (3.1) the notation is chosen so that the W_ρ^- contains annihilation operators of negatively charged particles W^- and, similarly, the W_ρ^+ involves annihilation operators of positively charged W^+ The coupling constant g is now dimensionless (similarly to spinor electrodynamics) as one can easily see from simple dimensional considerations (cf. Appendix G). The numerical factor $(2\sqrt{2})^{-1}$ in (3.1) is introduced as a commonly used convention.

3.2   Correspondence with Fermi Theory

The model of weak interactions defined by the Lagrangian (3.1) must respect an experimentally established fact that the effective Fermi-type theory (1.1) provides a good description of a considerable part of physical reality in the low-energy region. In the first place, this means that $W^{\pm}$ must have a non-zero mass (m_W), so that the model (3.1) would indeed describe short-range forces. (Let us remark that from negative results of direct search for $W^{\pm}$ it has long been known that if such a particle exists, it must be much heavier than e.g. the muon.) The condition of an equivalence of the IVB theory (3.1) and the Fermi-type theory (1.1) in the low-energy limit leads to a formula relating parameters G_F, g and m_W which will be repeatedly used in subsequent chapters. We will now derive this important relation.

Let us consider the muon decay $\mu \to e\nu_\mu\bar{\nu}_e$ as a typical example of a low-energy weak process. In the theory with IVB (3.1) such a process is described in lowest (i.e. 2nd) order of perturbation expansion by the Feynman diagram shown in Fig. 1(a), while in the Fermi-type theory the relevant diagram is that of Fig. 1(b) (here the lowest perturbative order means of course the 1st order in G_F).

The decay amplitude corresponding to the diagram 1(a) is given by the expression

$$
\begin{aligned}
i\mathcal{M}^{(a)}_{fi} \;=\;& i^3 \left(\frac{g}{2\sqrt{2}}\right)^2 [\bar{u}(k)\gamma_\rho(1-\gamma_5)u(P)]\,[\bar{u}(p)\gamma_\sigma(1-\gamma_5)v(k')] \times \\
& \times \frac{-g^{\rho\sigma} + m_W^{-2}q^\rho q^\sigma}{q^2 - m_W^2}
\end{aligned}
\tag{3.2}
$$

while the contribution of the graph 1(b) is

$$
i\mathcal{M}^{(b)}_{fi} = -i\frac{G_F}{\sqrt{2}}[\bar{u}(k)\gamma_\rho(1-\gamma_5)u(P)][\bar{u}(p)\gamma^\rho(1-\gamma_5)v(k')]
\tag{3.3}
$$

In Eq. (3.2) we have used the standard expression for the propagator of massive vector field (see (H.45) in Appendix H). Now we may let the second term in the numerator of the IVB propagator in Eq. (3.2) act on the matrix elements of fermion currents. Then using the Dirac equation for the corresponding spinors and taking into account the conservation of four-momentum $q = P - k = p + k'$ we obtain (assuming for simplicity that the neutrinos are massless)

$$
\begin{aligned}
\bar{u}(k)\slashed{q}(1-\gamma_5)u(P) &= m_\mu\bar{u}(k)(1+\gamma_5)u(P) \\
\bar{u}(p)\slashed{q}(1-\gamma_5)v(k') &= m_e\bar{u}(p)(1-\gamma_5)v(k')
\end{aligned}
\tag{3.4}
$$

From Eq. (3.4) it is clear that the contribution of the second term in the IVB propagator in Eq. (3.2) is suppressed by the factor $m_e m_\mu/m_W^2 \ll 1$ and thus it can be neglected. Further, simple kinematical considerations lead to the following bounds on the squared four-momentum of the virtual W in the diagram 1(b):

$$
m_e^2 \le q^2 \le m_\mu^2
\tag{3.5}
$$

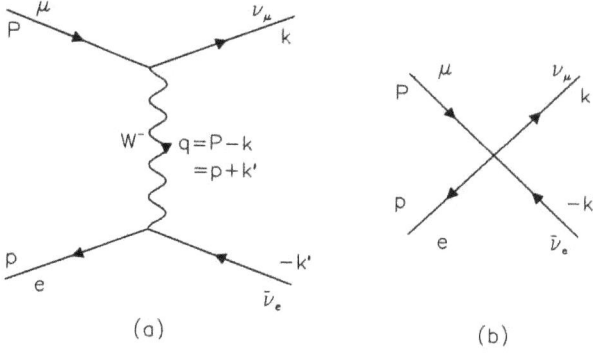

Fig. 1. *Feynman diagrams for the process* $\mu \to e\nu_\mu \bar{\nu}_e$ *(a) in the theory with IVB (b) in the Fermi-type theory.*

In view of what we have already said concerning the experimentally admissible value of m_W it is then also obvious from (3.5) that

$$q^2 \ll m_W^2 \tag{3.6}$$

so the momentum-dependence of the denominator of IVB propagator in Eq. (3.2) may be ignored. Comparing the expression (3.2) (in which the above-mentioned simplifications are taken into account) with (3.3) we get the desired relation

$$\frac{G_F}{\sqrt{2}} = \frac{g^2}{8m_W^2} \tag{3.7}$$

It is interesting to notice that in the derivation of Eq. (3.7) the negative sign in the Lagrangian of four-fermion interaction plays an important role; it is just this convention which then guarantees that $G_F > 0$, if the Fermi-type theory is viewed as an effective low-energy approximation of the theory with IVB.

3.3 Fermion Scattering Processes

We will now investigate the behaviour of scattering amplitudes and cross sections of processes $\nu_e e \to \nu_e e$ and $\bar{\nu}_e e \to \bar{\nu}_e e$ in the high-energy limit, i.e. for $s \gg m_W^2$ (for $s \ll m_W^2$ the effective Fermi-type theory is of course valid if the relation (3.7) is maintained). Feynman diagrams corresponding to these processes in the theory with IVB (3.1) (in tree approximation) are shown in Fig. 2.

Amplitudes corresponding to the diagrams in Fig. 2 are given by

$$i\mathcal{M}_{fi}^{(a)} = i^3 \left(\frac{g}{2\sqrt{2}}\right)^2 [\bar{u}(p')\gamma_\rho(1-\gamma_5)u(k)] [\bar{u}(k')\gamma_\sigma(1-\gamma_5)u(p)] \times$$

$$\times \ \frac{-g^{\rho\sigma} + m_W^{-2} q^\rho q^\sigma}{q^2 - m_W^2} \tag{3.8}$$

$$i\mathcal{M}_{fi}^{(b)} \ = \ i^3 \left(\frac{g}{2\sqrt{2}}\right)^2 [\bar{v}(k)\gamma_\rho(1-\gamma_5)u(p)]\,[\bar{u}(p')\gamma_\sigma(1-\gamma_5)v(k')] \times$$
$$\times \ \frac{-g^{\rho\sigma} + m_W^{-2} P^\rho P^\sigma}{P^2 - m_W^2} \tag{3.9}$$

Let us now try to estimate the high-energy behaviour of the expressions (3.8) and (3.9) with the help of dimensional considerations. In contrast with Fermi-type theory, the relevant coupling constant g is now dimensionless. However, the IVB propagator contains a term proportional to m_W^{-2}; thus, as the scattering amplitude $\mathcal{M}_{fi}$ is dimensionless, it might seem at first sight that it could grow linearly with s so as to compensate dimensionally the factor m_W^{-2}. In fact, the "dangerous" term in the IVB propagator in (3.8) or (3.9) may be eliminated by using the Dirac equation; lepton mass is factorized (cf. (3.4)) and instead of a term behaving as s/m_W^2 one gets a damping factor m_e^2/m_W^2. Thus, amplitudes (3.8) and (3.9) are asymptotically constant in the high-energy limit. More precisely, in the case of the expression (3.8) it is true for an arbitrary scattering angle different from 0 or π — this is obvious from the kinematical structure of the denominator of the corresponding propagator.

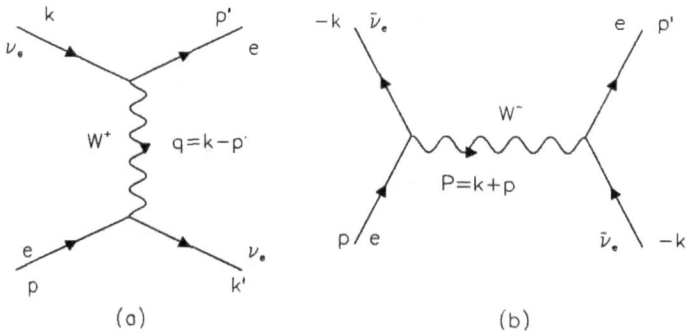

Fig.2. *Processes (a) $\nu_e e \to \nu_e e$ and (b) $\bar{\nu}_e e \to \bar{\nu}_e e$ in the second order of perturbation expansion in the theory with charged IVB. The relevant Mandelstam variables are $q^2 = u, P^2 = s$.*

In the high-energy limit (when one may set $m_e = 0$) the amplitude (3.8) is non-zero only for the combination of helicities $h_1 = h_2 = h_1' = h_2' = -\frac{1}{2}$ (cf. (2.6)); this is due to presence of the factor $1 - \gamma_5$ in charged weak currents. (In the case of the amplitude (3.9) the corresponding non-trivial combination is $h_1 = h'_1 = +\frac{1}{2}$, $h_2 = h'_2 = -\frac{1}{2}$, if we denote by h_1 and h'_1 helicities of the initial and final antineutrino.) Using the result (D.5) from Appendix D and repeating considerations similar to those which in the preceding chapter led to the relation (2.7), we obtain from Eq. (3.8) (for the above-mentioned combination of helicities and for $m_e = 0$)

$$
\begin{aligned}
|\mathcal{M}^{(a)}_{fi}| &= g^2 \frac{s}{|u - m_W^2|} \\
&= 2g^2 \frac{1}{1 + \cos\vartheta + 2m_W^2/s}
\end{aligned}
\tag{3.10}
$$

where ϑ is the neutrino scattering angle in the c.m. system. An exact (direct) calculation of the amplitude $\mathcal{M}^{(a)}_{fi}$ using explicit form of lepton spinors $u(p)$ (as given in Appendix B) recovers just the expression on the right-hand side of Eq. (3.10). This expression has (for any $\vartheta \neq \pi$) a finite limit for $s \to \infty$ (however, it behaves like s/m_W^2 for $\vartheta = \pi$).

The scattering amplitude for $\nu_e e \to \nu_e e$ given by Eq. (3.10) may now be expanded into partial waves. For the given combination of helicities we then have $\lambda = \lambda' = 0$ in the formula (E.6), i.e. we are dealing with an expansion into Legendre polynomials (see (F.4)). Amplitudes of partial waves may then be calculated by means of the formula (E.8). In the considered case the Jacob-Wick expansion involves an infinite number of partial waves owing to the dependence of the *denominator* in (3.10) on the angle ϑ (cf. the problem 2.5 at the end of Chapter 2). The lowest partial wave corresponds to $j = 0$. The formula (E.8) gives for the corresponding amplitude the result

$$
\begin{aligned}
\mathcal{M}^{(0)}(s) &= \frac{1}{32\pi} \int_{-1}^{1} \frac{2g^2}{1 + \cos\vartheta + 2m_W^2/s} d(\cos\vartheta) \\
&= \frac{g^2}{16\pi} \ln\left(\frac{s}{m_W^2} + 1\right)
\end{aligned}
\tag{3.11}
$$

Now imposing unitarity condition (2.5) on the partial-wave amplitude (3.11) we get (for $s/m_W^2 \gg 1$) the bound

$$
s \leq m_W^2 \exp\left(\frac{16\pi}{g^2}\right)
\tag{3.12}
$$

To assess a numerical value of the "unitarity bound" defined by the expression on the right-hand side of (3.12), let us e.g. assume that $g^2/4\pi \approx \alpha_{QED}$, where $\alpha_{QED} \approx \frac{1}{137}$ is the electromagnetic fine-structure constant. Then $16\pi/g^2 \approx 548$ and the unitarity condition (2.5) is violated only at astronomical energies, corresponding to $s \geq 10^{238} m_W^2$. (Let us remark that the present-day realistic value is about $g^2/4\pi \approx 0.032$; the right-hand side of (3.12) is then approximately equal to $10^{55} m_W^2$.)

In view of the functional form of the energy dependence of the partial-wave amplitude (3.11), such a case is usually referred to as a "logarithmic violation of unitarity" in tree approximation (note that a similar behaviour is also exhibited e.g. by partial-wave amplitudes in QED — see the problem 2.5 in previous chapter).

For completeness, let us also calculate cross sections corresponding to the amplitudes (3.8) and (3.9) in the asymptotic region $s \gg m_W^2$. Summing over lepton polarizations (and averaging with respect to the initial electron polarization) one gets (cf. (D.5), (D.6))

$$\overline{|\mathcal{M}_{fi}^{(a)}|^2} = \frac{1}{2}g^4 \frac{s^2}{(u - m_W^2)^2} \tag{3.13}$$

$$\overline{|\mathcal{M}_{fi}^{(b)}|^2} = \frac{1}{2}g^4 \frac{u^2}{(s - m_W^2)^2} \tag{3.14}$$

Employing the kinematical identity $u = -s(1 - y)$ (see (A.6)) and the formula (C.13) for differential cross section and performing finally an integration over y from 0 to 1, we obtain

$$\sigma_{IVB}^{(\nu e)} = \frac{G_F^2}{\pi} m_W^2 \frac{s}{s + m_W^2} \tag{3.15}$$

$$\sigma_{IVB}^{(\bar{\nu} e)} = \frac{G_F^2}{3\pi} m_W^4 \frac{s}{(s - m_W^2)^2} \tag{3.16}$$

To express the cross sections (3.15), (3.16) in terms of G_F, we have used the relation (3.7). Let us remark that while the result (3.15) represents a good approximation for an arbitrary $s \gg m_e^2$, the expression (3.16) may be used either for $s \gg m_W^2$ or $m_e^2 \ll s \ll m_W^2$; this of course is related to the fact that in the case of process $\bar{\nu}_e e \to \bar{\nu}_e e$ the W-exchange in the s-channel produces a pole in the corresponding propagator for $s = m_W^2$. This point will be mentioned briefly later in this chapter (see also the problem 3.3). From (3.15), (3.16) it is immediately seen that in the case of the neutrino process the corresponding cross section has a non-zero limit for $s \to \infty$

$$\sigma_{IVB}^{(\nu e)}\big|_{s \to \infty} = \frac{G_F^2}{\pi} m_W^2 \tag{3.17}$$

whereas the antineutrino cross section converges for $s \to \infty$ to zero as $1/s$:

$$\sigma_{IVB}^{(\bar{\nu} e)}\big|_{s \to \infty} \approx \frac{G_F^2}{3\pi} \frac{m_W^4}{s} \tag{3.18}$$

A technical remark may be in order here: Taking into account that both scattering amplitudes are asymptotically flat, a naive guess based on the formula (C.13) might be that both cross sections should vanish for $s \to \infty$. However, it is easy to see that the non-zero value in (3.17) is due to the fact that the amplitude for $\nu e \to \nu e$ is asymptotically bounded by a constant for all directions *except* $\vartheta = \pi$ (see Eq. (3.10)); note also that the same feature of (3.10) is responsible for the logarithmic growth of partial-wave amplitudes (cf. Eq. (3.11)).

Preceding considerations concerning the high-energy behaviour of the amplitudes of physical scattering processes in the IVB theory (3.1) may be summarized briefly as follows: From the technical point of view, the idea of massive charged IVB as an "agent" of weak interactions seems to be somewhat problematic at first sight because of the longitudinal piece of the vector boson propagator involving the factor m_W^{-2} which could, in principle, play the same role as the coupling constant G_F in the Fermi-type theory. Nevertheless, an application of the equations of motion (i.e. the Dirac equation) eliminates potential problems at least in the case of purely fermionic processes. The corresponding tree-level scattering amplitudes are asymptotically flat in high-energy limit and a violation of unitarity is described at worst by a logarithmic function of energy (contrary to the power-like growth of partial-wave tree amplitudes in Fermi-type theory).

3.4 Process $\nu\bar{\nu} \to W_L^- W_L^+$

The model with charged IVB thus represents in a sense a more satisfactory theoretical description (from the technical point of view) of purely fermionic scattering processes than the Fermi – Feynman – Gell-Mann model (1.1). However, this success is far from complete. Since we have introduced IVB as a new object into the theory of weak interactions, it is natural to consider a direct production of physical $W^\pm$ as well as processes involving a virtual IVB. In doing this, it turns out that for some combinations of polarizations of external $W^\pm$ the amplitudes of the corresponding (tree-level) diagrams exhibit a power-like growth in high-energy limit. A classic example of such a process is the production of a pair of $W^\pm$ in the neutrino - antineutrino annihilation, i.e.

$$\nu\bar{\nu} \to W^- W^+ \tag{3.19}$$

(In what follows, unless stated otherwise, we are working with electron-type leptons and the corresponding index e is systematically omitted.) The process (3.19) was first discussed in this context in the paper [24]. (It is a certain historical paradox that the paper [24] appeared only 2 years after Weinberg's work [7] and that Weinberg's paper is not even mentioned in [24]. In contrast to the commonly accepted notation the authors of [24] use a symbol X for the charged IVB.) We will now derive the essential properties of the tree-level amplitude of the process (3.19) in the high-energy limit. The corresponding lowest-order Feynman diagram is shown in Fig. 3.

First of all, one has to realize that a possible source of "bad" high-energy behaviour of the diagram in Fig. 3 (i.e. a power-like growth of the corresponding amplitude with energy) may reside in polarization vectors of the final-state $W^\pm$. Indeed, components of the vector of *longitudinal* polarization (corresponding to zero helicity) grow linearly with energy in the ultrarelativistic limit (see Appendix H, Eq. (H.25)):

$$\varepsilon_L^\mu(p) = \frac{1}{m_W} p^\mu + O(\frac{m_W}{p_0}) \tag{3.20}$$

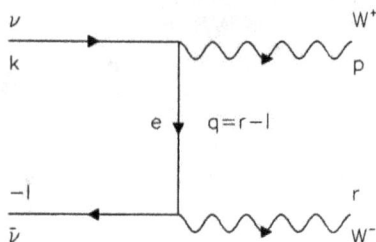

Fig.3. The process $\nu\bar{\nu} \to W^+W^-$ in the second order of perturbation expansion in the theory with charged IVB.

(Let us however stress that the normalization $\varepsilon_L.\varepsilon_L^* = -1$ is still maintained!) The leading term in the longitudinal polarization (i.e. the first term in (3.20)) is thus proportional to the corresponding four-momentum; the presence of the factor m_W^{-1} in this term will always play a key role in the estimates of the high-energy asymptotics of tree-level amplitudes for processes involving real (i.e. physical) massive vector bosons, both here and in the subsequent chapters. Let us now consider the contribution of the diagram in Fig. 3 when both final-state W's have longitudinal polarizations; in such a case one may expect the worst behaviour of the corresponding scattering amplitude in the high-energy limit. The character of the leading divergence for $s \to \infty$ may easily be guessed: Taken together, the leading terms from $\varepsilon_L(p)$ and $\varepsilon_L(r)$ produce, according to Eq. (3.20), a factor of m_W^{-2} and for dimensional reasons (scattering amplitude of a binary process must be dimensionless) one may thus expect a quadratic dependence on energy for the leading term in the considered amplitude. Further, it is also obvious that it is just the combination of leading terms in both longitudinal polarizations which may yield expressions divergent for $s \to \infty$; all the other combinations may only contribute to the asymptotically constant (i.e. $O(1)$) terms in the limit $s \to \infty$. Taking into account the above remarks, the amplitude for the process $\nu\bar{\nu} \to W_L^+W_L^-$ corresponding to the diagram in Fig. 3 may be expressed as

$$
\begin{aligned}
i\mathcal{M}_{fi} &= i^3 (\frac{g}{2\sqrt{2}})^2 \bar{v}(l)\gamma_\mu(1-\gamma_5)\frac{1}{\not{q}-m}\gamma_\nu(1-\gamma_5)u(k)\varepsilon_L^{*\mu}(r)\varepsilon_L^{*\nu}(p) = \\
&= -i\frac{g^2}{8}\bar{v}(l)\gamma_\mu(1-\gamma_5)\frac{1}{\not{q}-m}\gamma_\nu(1-\gamma_5)u(k)\frac{r^\mu}{m_W}\frac{p^\nu}{m_W} + O(1)
\end{aligned}
$$

$$(3.21)$$

(the standard form of the electron propagator used in Eq. (3.21) of course represents the inverse matrix $(\not{q}-m)^{-1}$; one should keep this in mind in subsequent manipulations).

The relation (3.21) may be further rewritten in the following way: We employ the energy-momentum conservation $q = r - l$ (see Fig. 3), decompose "artificially" the $\not{r}$ as $\not{r} = \not{r} - \not{l} + \not{l} = \not{q} + \not{l}$ and use the Dirac equation $\bar{v}(l)\not{l} = 0$ (we of course assume that $m_\nu = 0$). Then we obtain, after a simple algebraic manipulation

$$\mathcal{M}_{fi} = -\frac{g^2}{8m_W^2}\bar{v}(l)(1 + \gamma_5)\not{q}\frac{1}{\not{q} - m}\not{p}(1 - \gamma_5)u(k) + O(1) \tag{3.22}$$

In the last expression one may again use an artificial decomposition $\not{q} = \not{q} - m + m$; by means of this simple trick and performing some additional standard manipulations we recast Eq. (3.22) as

$$
\begin{aligned}
\mathcal{M}_{fi} = & -\frac{g^2}{4m_W^2}\bar{v}(l)\not{p}(1 - \gamma_5)u(k) \\
& -\frac{g^2}{8m_W^2}m\bar{v}(l)(1 + \gamma_5)\frac{\not{q} + m}{q^2 - m^2}\not{p}(1 - \gamma_5)u(k) \\
& + O(1)
\end{aligned}
\tag{3.23}
$$

The first term on the right-hand side of Eq. (3.23), i.e.

$$\mathcal{M}_{fi}^{(1)} = -\frac{g^2}{4m_W^2}\bar{v}(l)\not{p}(1 - \gamma_5)u(k) \tag{3.24}$$

is, as expected, quadratically divergent for $E_{c.m.} \to \infty$ (let us recall that lepton spinors $u(k), v(l)$ behave in the high-energy limit like $E_{c.m.}^{1/2}$ (i.e. $s^{1/4}$) for the chosen normalization). In the terminology which we will use in what follows, the term (3.24) represents the leading (or dominant) divergence of the considered tree-level amplitude. For a more detailed representation of this leading term as an explicit function of energy we refer the reader e.g. to the textbook [25] or the original paper [24]. However, we will not need such detailed formulae; expressions of the type (3.24) will be sufficient for our purposes.

We will now examine the second term on the right-hand side of Eq. (3.23). One might expect *a priori* that this expression contains a next-to-leading (in this case linear) divergence for $E_{c.m.} \to \infty$. However, the would-be linear divergence can easily be seen to vanish identically since

$$(1 + \gamma_5)\not{q}\not{q}\not{p}(1 - \gamma_5) = 0$$

Thus, in the second term on the right-hand side of Eq. (3.23) the electron mass squared m^2 is in fact factorized, which compensates the coefficient m_W^{-2} coming from longitudinal polarizations and the whole expression is therefore of the order $O(1)$ for $s \to \infty$. From the calculation that we have just described it is also clear that the elimination of the linearly divergent term is a consequence of the assumption $m_\nu = 0$, more precisely of the fact that the initial-state fermions (i.e. $\nu, \bar{\nu}$) are massless - for an illustration see also the problem 3.6 at the end of this chapter. (Such a connection

will play an important role in the derivation of the standard model in Chapter 5.)
For the tree-level amplitude of the process $\nu\bar{\nu} \to W_L^+ W_L^-$ we thus have the result

$$\mathcal{M}_{fi} = \mathcal{M}_{fi}^{(1)} + O(1) \tag{3.25}$$

where the leading term $\mathcal{M}_{fi}^{(1)}$ is given by the formula (3.24). Quadratic growth of
this term with energy means that perturbative S-matrix unitarity for the considered
process is violated in the same way as it was the case for four-fermion scattering
processes in the Fermi-type theory.

Some information concerning the behaviour of the considered model in higher
orders of perturbation expansion is contained in the "effective index" of the corresponding interaction vertex which is defined and calculated in Appendix G (see the
formula (G.14)):

$$\omega_v^{eff.} = \frac{3}{2}n_F + 2n_B + n_D$$

(Let us recall that the coefficient 2 multiplying the number of boson lines n_B involved
in the interaction vertex in this case is a consequence of the ultraviolet behaviour of the
canonical massive vector boson propagator.) In our case $n_F = 2$, $n_B = 1$ and $n_D = 0$,
so

$$\omega_v^{eff} = 5$$

(let us remind the reader that in the Fermi-type theory one has $\omega_v = 6$). The
value $\omega_v^{eff} = 5 > 4$ indicates non-renormalizability in higher orders of perturbation
expansion, and a detailed analysis has indeed led to the conclusion that the model
of weak interactions described by the Lagrangian (3.1) is not renormalizable within
the framework of perturbation expansion (see [26], [27]). However, the following
remark is in order here: One has to keep in mind that the inequality $\omega_v^{eff} \leq 4$ is
in general not a *necessary* condition of perturbative renormalizability for a quantum
field theory model. For example, in massive QED one also has $\omega_v^{eff} = 5$, but this
theory is still renormalizable as we have already stressed in the preceding chapter.
Another important example of a theory which violates the condition $\omega_v^{eff} \leq 4$ but
nevertheless produces a renormalizable perturbation expansion for the S-matrix is
just the standard GWS model.

As we have seen, the theory of weak interactions with charged IVB is non-
renormalizable and some scattering amplitudes corresponding to tree diagrams diverge severely in the high-energy limit (displaying a power-law behaviour). Thus, as
in the case of the Fermi-type theory one may observe here a remarkable connection
between two different aspects of the perturbation expansion mentioned at the end of
Chapter 2: The power-like growth of tree-level amplitudes in the high-energy region
(for real particles) implies non-renormalizability at higher orders of the perturbation
expansion, i.e. an unacceptable behaviour of Feynman diagrams in the ultraviolet
domain of four-momenta (of virtual particles) in closed loops of internal lines. In the
next chapter we will examine the electrodynamics of charged massive vector bosons
from this point of view.

3.5   Lepton Decays of the IVB

To close this chapter we shall now discuss briefly lepton decays of the IVB. The processes we have considered in the IVB theory up to now corresponded to diagrams of at least second order in perturbation expansion. However, the theory described by the interaction Lagrangian (3.1) also admits (for a sufficiently heavy IVB) processes of decay of $W^\pm$ into a lepton pair; the corresponding decay amplitude is non-zero already in the first order of perturbation expansion (i.e. in the first order of g). According to our conventions the first term in the Lagrangian (3.1) describes the decay $W^+ \to e^+ + \nu$ while the second term yields

$$W^- \to e^- + \bar\nu \qquad (3.26)$$

For definiteness we shall deal with the process (3.26). The corresponding tree-level Feynman diagram is shown in Fig. 4.

The probability of the decay per unit time, i.e. the decay rate (or width) corresponding to the process (3.26) may be calculated by means of the formula (C.19) from Appendix C (we assume that all particles are unpolarized). For simplicity we will also neglect the electron mass m; taking into account that $m \ll m_W$, it is clear that such a simplification is in fact a very good approximation. A detailed calculation for $m \neq 0$ may be left to the interested reader as an instructive exercise (see the problem 3.8).

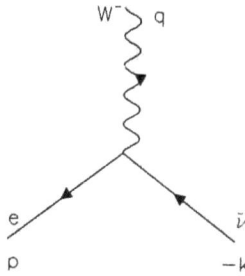

Fig.4. The process $W^- \to e\bar\nu$ in the lowest order of perturbation expansion.

Before performing the formal calculation it is useful to realize that in the approximation $m = 0$ one may easily guess the dependence of the considered decay width on the other relevant physical parameters, i.e. on g and m_W: The decay width has the dimension of a mass in our system of units; the only mass which is now available is the m_W and one must therefore have $\Gamma \sim m_W$. Further, the decay amplitude is proportional to g (in the first perturbative order) and thus $\Gamma \sim g^2$ The considered decay rate must therefore necessarily have (for $m = 0$) the form

$$\Gamma(W^- \to e\bar{\nu}) = C g^2 m_W \qquad (3.27)$$

where C is a numerical constant.

. We will now determine this constant by means of an explicit calculation. The contribution of the diagram in Fig. 4 is given by the expression

$$\mathcal{M}_{fi} = \frac{g}{2\sqrt{2}}\bar{u}(p)\gamma_\rho(1 - \gamma_5)v(k)\varepsilon^\rho(q) \qquad (3.28)$$

where $\varepsilon^\rho(q)$ is the polarization vector of IVB; of course, $q = k + p$. The calculation of the squared modulus of the invariant amplitude (3.28) and the summation over polarizations may most effectively be carried out in the following way: First we sum over polarizations of the decaying IVB by means of the formula (H.28) (see Appendix H) to get

$$\sum_{pol.} |\mathcal{M}_{fi}|^2 = \frac{g^2}{8} \sum_{lepton\ pol.} [\bar{u}(p)\gamma_\rho(1 - \gamma_5)v(k)] \times$$
$$\times \ [\bar{v}(k)\gamma_\sigma(1 - \gamma_5)u(p)]\left(-g^{\rho\sigma} + \frac{1}{m_W^2}q^\rho q^\sigma\right) \qquad (3.29)$$

However, the term involving $m_W^{-2}q^\rho q^\sigma$ gives zero contribution; this is immediately obvious if we use the Dirac equation (for $m = 0$ both vector and axial-vector current are exactly conserved). From Eq. (3.29) it then easily follows that

$$\sum_{pol.} |\mathcal{M}_{fi}|^2 = -\frac{1}{4}g^2 Tr[\not{p}\gamma_\rho \not{k}\gamma^\rho(1 - \gamma_5)] =$$
$$= 2g^2(k.p) = g^2 m_W^2 \qquad (3.30)$$

Averaging over the vector boson polarizations amounts to multiplying Eq. (3.30) by a factor of $\frac{1}{3}$. Then using formulae (C.19) and (C.22) we get finally

$$\Gamma(W^- \to e\bar{\nu}) = \frac{1}{48\pi}g^2 m_W \qquad (3.31)$$

The coefficient C in our preliminary estimate (3.27) is thus seen to be $(48\pi)^{-1}$. For the rate of the charge-conjugate process $W^+ \to e^+\nu$ we of course get the same result. Using the relation (3.7) the result (3.31) may be recast as

$$\Gamma(W^- \to e\bar{\nu}) = \frac{1}{6\pi\sqrt{2}}G_F m_W^3 \qquad (3.32)$$

Let us remark that the above calculation is not just an academic exercise within the framework of a provisional theory of weak interactions; the Lagrangian (3.1) in fact makes *a part* of the GWS standard model and the result (3.31) or (3.32) thus holds (in lowest order) without any change even in the modern theory of electroweak interactions. Note finally that we could also take into account the hadronic part of the weak current in the Lagrangian (3.1) and calculate the corresponding decay rate for hadron (i.e. quark) modes. We defer such a calculation to the last chapter devoted to the standard model where we also discuss the slightly more complicated pattern of mixing in the quark sector which seems to occur in the real world (see the problem 5.18 at the end of Chapter 5).

For the current experimental value $m_W \doteq 80.2 \text{GeV}$ (see [28]) the decay rate for the electronic mode (3.32) is

$$\Gamma(W^- \to e\bar{\nu}) \doteq 230 \text{MeV} \tag{3.33}$$

The value of this partial width thus shows that the mean lifetime of the charged IVB is shorter than $10^{-23} sec$ (which is a typical lifetime of hadron resonances, e.g. the meson $\rho(770)$).

Coming back to the relations (3.14) or (3.16), we see that according to this theory, the intermediate vector boson should manifest itself as a dramatic enhancement of the scattering cross section for $\bar{\nu}e \to \bar{\nu}e$ in the vicinity of $s = m_W^2$ (an experimental verification of this undoubtedly correct prediction will, however, be out of reach of terrestrial facilities in the foreseeable future). The instability of the IVB (i.e. its finite decay width Γ) leads to a modification of the denominator of the corresponding propagator: The standard Feynman expression

$$q^2 - m_W^2 + i\varepsilon \tag{3.34}$$

(corresponding to a stable particle) turns into a "Breit-Wigner form"

$$q^2 - m_W^2 + im_W\Gamma \tag{3.35}$$

Let us remark that in the GWS theory the passage from (3.34) to (3.35) may be formally accomplished by including higher-order effects, i.e. perturbative corrections to the propagator on the level of diagrams with (at least) one closed loop (see e.g. [39]). The modification (3.35) obviously regulates the original singularity (pole) in the IVB propagator, which would appear in scattering amplitude of the process $\bar{\nu}e \to \bar{\nu}e$ for $s = m_W^2$ (cf. (3.16)). As we have already observed, the corresponding cross section should rather display resonance behaviour with a maximum at $s = m_W^2$ (in this context, see also the problem 3.3).

Problems

3.1. In the theory with charged IVB calculate the cross section of the process $e^-e^+ \to \nu\bar{\nu}$ (in the tree approximation) in the limit $s \gg m_e^2$, i.e. effectively for $m_e = 0$.

Compare the result with the cross section of the process $e^- e^+ \to \mu^- \mu^+$ in QED for $s \gg m_\mu^2$ (see (D.18) in Appendix D).

3.2. Calculate amplitudes of the partial waves with $j = 1$ and $j = 2$ (in the tree approximation) for the process $\nu e \to \nu e$ (set $m_e = 0$). Show that partial-wave amplitudes for an arbitrary j grow logarithmically with energy.

3.3. How many partial waves contribute to the Jacob-Wick expansion of the scattering amplitude for the process $\bar{\nu} e \to \bar{\nu} e$? Calculate the corresponding partial-wave amplitudes and the cross section (again for $m_e = 0$); take into account the effect of the finite width of W. What restriction is imposed by unitarity in this case?

3.4. Examine the asymptotic behaviour of the tree-level amplitude for the process $\nu \bar{\nu} \to W_L^- W_T^+$, where the indices L and T denote the longitudinal and transverse polarization respectively.

3.5. Calculate the leading term in the cross section $\sigma(\nu \bar{\nu} \to W^- W^+)$ for unpolarized $W^\pm$ in the high-energy limit.

3.6. Examine the high-energy behaviour of the tree diagram corresponding to the process $e^- e^+ \to W_L^- W_L^+$ in the theory described by the Lagrangian (3.1). Calculate also the leading asymptotic term in the corresponding cross section for unpolarized particles.

3.7. Consider the process $e^- e^+ \to \gamma\gamma$ in the case that the photon mass is different from zero. What is the high-energy behaviour of the corresponding tree-level amplitude for longitudinally polarized "heavy photons"?

3.8. Calculate the decay width $\Gamma(W^- \to e\bar{\nu})$ for $m_e \neq 0$.

3.9. Calculate the decay width $\Gamma(e^- \to W^- + \nu_e)$ in a hypothetical world where $m_e > m_W$ (and $m_\nu = 0$). (Note that this rather academic example is a prototype of the realistic process $t \to W^+ + b$, where t, b are quarks from the third generation of fermions in the framework of the standard model.)

Chapter 4

ELECTRODYNAMICS OF VECTOR BOSONS

4.1   Interactions of $W^{\pm}$ with Photons

The intermediate vector boson of weak interactions carries an electric charge (as it is coupled to a charged fermionic current) and it is therefore natural to consider also electromagnetic interactions of the particles $W^{\pm}$. Electrodynamics of charged IVB is the subject of this chapter. As we will see, in contrast with the familiar "textbook" spinor electrodynamics (where the charged particles have spin $\frac{1}{2}$) the electrodynamics of massive vector bosons (i.e. charged spin-1 particles) is non-renormalizable within the perturbative framework. More precisely, we will show here that amplitudes of some tree diagrams in this theory display an equally bad high-energy behaviour (i.e. a power-like growth) as that we have observed in the model of weak interactions described in the preceding chapter. The non-renormalizability at higher orders of perturbation expansion has been demonstrated in [26]. Electrodynamics of charged massive vector bosons has been discussed in many papers published in the 1960's (see e.g. [29-32] and other papers quoted therein); cf. also [18], [33] and for a recent reference see in particular [34].

An electromagnetic interaction of the IVB may be introduced (similarly to the case of charged spin -$\frac{1}{2}$ fermions) by means of a suitable gauge-invariant modification of the corresponding free Lagrangian. The Lagrangian of free (non-interacting) fields $W^{\pm}$ is given by (see (H.47) in Appendix H)

$$\mathcal{L}_0 = -\frac{1}{2}(\partial_\mu W_\nu^- - \partial_\nu W_\mu^-)(\partial^\mu W^{+\nu} - \partial^\nu W^{+\mu}) + m_W^2 W_\mu^- W^{+\mu} \qquad (4.1)$$

The "minimal" electromagnetic interaction is defined by changing (4.1) into

$$\mathcal{L}_{EM}^{(min.)} = -\frac{1}{2}(D_\mu W_\nu^- - D_\nu W_\mu^-)(D^{\mu*} W^{+\nu} - D^{\nu*} W^{+\mu}) + m_W^2 W_\mu^- W^{+\mu} \qquad (4.2)$$

where D_μ is the covariant derivative

$$\begin{aligned} D_\mu &= \partial_\mu + ieA_\mu \\ D_\mu^* &= \partial_\mu - ieA_\mu \end{aligned} \qquad (4.3)$$

(the coupling constant in (4.3) is $e > 0$). The Lagrangian (4.2) is invariant under local gauge transformations

$$
\begin{aligned}
W_\mu^{-\prime}(x) &= e^{-i\omega(x)} W_\mu^-(x) \\
W_\mu^{+\prime}(x) &= e^{+i\omega(x)} W_\mu^+(x) \\
A_\mu'(x) &= A_\mu(x) + \frac{1}{e}\partial_\mu \omega(x)
\end{aligned}
\tag{4.4}
$$

Let us emphasize that gauge transformations (4.4) correspond, as in the spinor electrodynamics, to an abelian (i.e. commutative) group $U(1)$.

One may add to the "minimal" Lagrangian (4.2) another gauge invariant term

$$
\mathcal{L}' = -i\kappa e W^{-\mu} W^{+\nu} F_{\mu\nu}
\tag{4.5}
$$

where

$$
F_{\mu\nu} = \partial_\mu A_\nu - \partial_\nu A_\mu
\tag{4.6}
$$

and κ is an arbitrary (real) constant. If we require a general electromagnetic interaction to be described only by polynomials with canonical dimension not greater than four (so as not to spoil renormalizability *a priori*) and, moreover, if we assume the invariance with respect to discrete symmetries C, P, T (for a more detailed discussion see e.g. [34]), then the most general Lagrangian of electrodynamics of the spin-1 charged vector bosons $W^\pm$ is obtained by summing (4.2) and (4.5):

$$
\mathcal{L}_{EM} = \mathcal{L}_{EM}^{(min.)} + \mathcal{L}' = \mathcal{L}_0 + \mathcal{L}_{int}^{(min.)} + \mathcal{L}'
\tag{4.7}
$$

An alternative (and in a sense more general) approach to electromagnetic interactions of $W^\pm$ is discussed in Appendix I and in Chapter 5 (see Section 5.4). Let us remark that adding the term (4.5) to the original minimal interaction incorporated in (4.2) corresponds physically to particles $W^\pm$ with an "anomalous" magnetic moment $\mu_W = (1+\kappa)e/(2m_W)$ (the corresponding gyromagnetic factor is thus $g = 1+\kappa$) and electric quadrupole moment $Q_W = \kappa e m_W^{-2}$ (see e.g. [17], p. 22 and also the papers [33], [34]). Let us recall that the gyromagnetic factor $g = 2$ for *electron* follows automatically from the Dirac equation with *minimal* electromagnetic interaction, while in the case of vector bosons the value of $g = 2$ corresponds to $\kappa = 1$ in (4.5). It is also useful to realize that both the minimal interaction $\mathcal{L}_{int}^{(min.)}$ and the term $\mathcal{L}'$ in (4.7) have the same canonical dimension (equal to four) and thus there is no reason to prefer *a priori* any particular value of the parameter κ; in this context, instead of "anomalous", perhaps a more correct adjective "ambiguous" is used for the magnetic moment of $W^\pm$ (see e.g. [18]). In spinor electrodynamics, an analogue of the non-minimal term (4.5) is the expression $\bar{\psi}\sigma_{\mu\nu}\psi F^{\mu\nu}$, which, however, has dimension 5 and would lead to a non-renormalizable perturbation expansion.

Using (4.2), (4.3), (4.5) and (4.6) we may recast the interaction part of the Lagrangian (4.7) as

$$
\mathcal{L}_{int} = \mathcal{L}_{int}^{(min.)} + \mathcal{L}' = \mathcal{L}_{WW\gamma} + \mathcal{L}_{WW\gamma\gamma}
\tag{4.8}
$$

where for the term trilinear with respect to the fields $W^{\pm}$ and A_μ (photon) one gets, after a straightforward manipulation

$$
\begin{aligned}
\mathcal{L}_{WW\gamma} = \; & - \; ie[A^\mu(W^{-\nu}\partial_\mu W_\nu^+ - \partial_\mu W_\nu^- W^{+\nu}) \\
& + \; W^{-\mu}(\kappa W^{+\nu}\partial_\mu A_\nu - \partial_\mu W^{+\nu} A_\nu) \\
& + \; W^{+\mu}(A^\nu \partial_\mu W_\nu^- - \kappa \partial_\mu A^\nu W_\nu^-)]
\end{aligned}
\tag{4.9}
$$

and the quadrilinear term is given by

$$
\mathcal{L}_{WW\gamma\gamma} = -e^2(A_\mu A^\mu W_\nu^- W^{+\nu} - A^\mu A^\nu W_\mu^- W_\nu^+)
\tag{4.10}
$$

As we have already said, the value $\kappa = 0$ in (4.9) corresponds to the minimal electromagnetic interaction. In what follows, the particular case $\kappa = 1$ will play the most important role; the corresponding trilinear interaction (4.9) will be called the electromagnetic interaction of the Yang-Mills type and denoted as $\mathcal{L}_{WW\gamma}^{(YM)}$ because in such a case, the expression (4.9) just corresponds to the situation where $W_\mu^{\pm}$ and A_μ form a triplet of non-abelian gauge (i.e. Yang-Mills) fields (see [8] and [17], [18], [25] etc.). The expression (4.9) is remarkably symmetric for $\kappa = 1$ (it is invariant with respect to cyclic permutations of W^-, W^+ and A) and may be recast in the more compact form:

$$
\mathcal{L}_{WW\gamma}^{(YM)} = -ie(A^\mu W^{-\nu}\overleftrightarrow{\partial}_\mu W_\nu^+ + W^{-\mu}W^{+\nu}\overleftrightarrow{\partial}_\mu A_\nu + W^{+\mu}A^\nu\overleftrightarrow{\partial}_\mu W_\nu^-)
\tag{4.11}
$$

The symbol $\overleftrightarrow{\partial}$ in (4.11) is defined in the usual way as

$$
f\overleftrightarrow{\partial}_\mu g = f(\partial_\mu g) - (\partial_\mu f)g
$$

Interaction vertices corresponding to the Lagrangian (4.9), (4.10) in momentum representation are shown in Fig. 5. The vertex in Fig. 5(a) corresponds to the expression

$$
\mathcal{V}_{\lambda\mu\nu}^{(\gamma)}(k,p,q|\kappa) = eV_{\lambda\mu\nu}(k,p,q|\kappa)
\tag{4.12}
$$

where

$$
V_{\lambda\mu\nu}(k,p,q|\kappa) = (k-p)_\nu g_{\lambda\mu} + (p-\kappa q)_\lambda g_{\mu\nu} + (\kappa q - k)_\mu g_{\lambda\nu}
\tag{4.13}
$$

One has to keep in mind the four-momentum conservation in (4.13)

$$
k + p + q = 0
\tag{4.14}
$$

In Feynman diagrams involving vertices of the type $WW\gamma$ an incoming line of the W^- with a four-momentum k is equivalent to an outgoing line of the W^+ with four-momentum $-k$ etc. In the case of interaction of the Yang-Mills type (i.e. for $\kappa = 1$) we will write simply

$$
V_{\lambda\mu\nu}(k,p,q) = (k-p)_\nu g_{\lambda\mu} + (p-q)_\lambda g_{\mu\nu} + (q-k)_\mu g_{\lambda\nu}
\tag{4.15}
$$

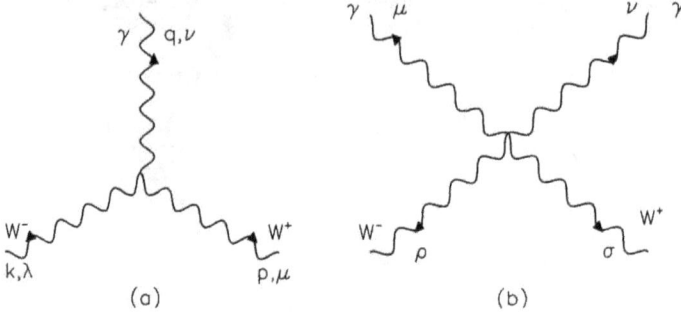

Fig.5. (a) Vertex $W^-W^+\gamma$ corresponding to the trilinear interaction (4.9); (b) Vertex $W^-W^+\gamma\gamma$ corresponding to (4.10).

The general vertex $WW\gamma$ (4.13) may then be expressed as

$$V_{\lambda\mu\nu}(k,p,q|\kappa) = V_{\lambda\mu\nu}(k,p,q) + (1-\kappa)(q_\lambda g_{\mu\nu} - q_\mu g_{\lambda\nu}) \tag{4.16}$$

The vertex $WW\gamma\gamma$ in Fig. 5(b) is given by

$$\mathcal{V}_{\mu\nu\rho\sigma} = -e^2(2g_{\mu\nu}g_{\rho\sigma} - g_{\mu\rho}g_{\nu\sigma} - g_{\mu\sigma}g_{\nu\rho}) \tag{4.17}$$

(in contrast to (4.13) the last expression is momentum-independent as the interaction (4.10) does not involve derivatives).

The Yang-Mills expression (4.15) is invariant with respect to a simultaneous cyclic permutation of the indices λ, μ, ν and of the momenta k, p, q, i.e.,

$$V_{\lambda\mu\nu}(k,p,q) = V_{\mu\nu\lambda}(p,q,k) = V_{\nu\lambda\mu}(q,k,p) \tag{4.18}$$

and it satisfies an important relation

$$p^\mu V_{\lambda\mu\nu}(k,p,q) = (-q^2 g_{\lambda\nu} + q_\lambda q_\nu) - (-k^2 g_{\lambda\nu} + k_\lambda k_\nu) \tag{4.19}$$

The relation (4.19) is sometimes called the 't Hooft identity since it probably first appeared in the paper [35]. The proof is left to the reader as an easy exercise (see the problem 4.1).

4.2   High-Energy Behaviour and the Vertex $WW\gamma$

We will examine the high-energy behaviour of tree-level scattering amplitudes of electromagnetic processes involving real (physical) vector bosons $W^\pm$ in the initial

and/or final state. (Throughout the following text, we will employ the generic notation E for a relevant energy, e.g. $E = E_{c.m.} = \sqrt{s}$.) For a discussion of different variants of the trilinear interaction $WW\gamma$ in (4.9), the most interesting processes in this context are those involving two $W's$ and two photons, i.e. for example
$W^-W^+ \to \gamma\gamma$, $W^-\gamma \to W^-\gamma$ etc., since the corresponding Feynman diagrams contain both external and internal W lines. For definiteness, let us first consider the
annihilation of a $W^\pm$ pair into two photons. The diagrams corresponding to this
process in the lowest order of perturbation expansion (i.e. in the 2nd order with respect to the interaction (4.9) and in the 1st order with respect to (4.10)) are depicted
in Fig. 6. On the basis of simple considerations similar to those employed in the

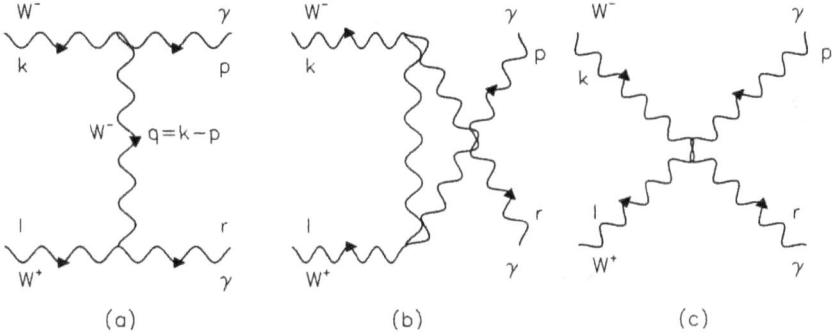

Fig.6. *Tree-level diagrams for the process* $W^-W^+ \to \gamma\gamma$ *contributing to order* e^2 *in
the electromagnetic coupling constant.*

preceding chapter one may guess that the high-energy behaviour of the diagrams (a),
(b) will in general be worse than in the case (c). The reason for this is of course the
factor of m_W^{-2} in the W propagator; in principle, it could cause contributions of the
diagrams (a), (b) to grow as E^{n+2} for $E \to \infty$, while the diagram (c) behaves like
E^n for some $n \geq 0$ (according to (H.25) or (3.20), further factors of m_W^{-1} may arise
from longitudinal polarizations of the external $W^\pm$, but these are common for all the
diagrams (a), (b), (c)). Of course, such a behaviour of the diagrams (a), (b) would
disqualify the electrodynamics of vector bosons $W^\pm$ *a priori*, as the above-mentioned
leading divergences - if they are present - remain unmatched.

A *necessary* condition for the absence of a power-like growth of the considered tree-
level amplitude with energy is therefore an elimination of the leading divergences in
the diagrams (a), (b) themselves. Since a general $WW\gamma$ interaction defined by (4.9)
(or (4.16)) depends on an arbitrary parameter κ, one may try to achieve the desired
divergence cancellation by means of an appropriate choice of the κ. To see how this
can be done, we will investigate the diagram (a) in detail (the diagram (b) behaves

analogously). Its contribution may be written as a sum of two expressions which correspond to the two terms in the W propagator (cf. (H.45)):

$$\mathcal{M}_a = \mathcal{M}_a^{(1)} + \mathcal{M}_a^{(2)} \tag{4.20}$$

where $\mathcal{M}_a^{(1)}$ corresponds to the diagonal term of the propagator and $\mathcal{M}_a^{(2)}$ contains the factor m_W^{-2} (it corresponds to the longitudinal term). In view of what we have already said, it is just the second term which is essential for our discussion. The expression $\mathcal{M}_a^{(2)}$ is given by

$$\mathcal{M}_a^{(2)} = -\frac{e^2}{m_W^2}\frac{q^\mu q^\nu}{q^2 - m_W^2}V_{\sigma\mu\lambda}(p, q, -k|\kappa)V_{\rho\tau\nu}(r, -l, -q|\kappa)\varepsilon^\lambda(k)\varepsilon^\tau(l)\varepsilon^{*\sigma}(p)\varepsilon^{*\rho}(r) \tag{4.21}$$

(the term $\mathcal{M}_a^{(1)}$ in (4.20) may be obtained from (4.21) by replacing $m_W^{-2}q^\mu q^\nu$ with $-g^{\mu\nu}$ and in the high-energy limit it behaves similarly to the diagram (c)). To work out the expression (4.21) we use the relations (4.16), (4.18), the 't Hooft identity (4.19) and simple kinematical relations $q = k - p = r - l$, $k^2 = l^2 = m_W^2$, $p^2 = r^2 = 0$. Thus we get first

$$\begin{aligned}
\mathcal{M}_a^{(2)} &= -\frac{e^2}{m_W^2}\frac{1}{q^2 - m_W^2}\varepsilon^\lambda(k)\varepsilon^\tau(l)\varepsilon^{*\sigma}(p)\varepsilon^{*\rho}(r) \times \\
&\times [k_\lambda k_\sigma - p_\lambda p_\sigma + (1 - \kappa)(k.p\, g_{\lambda\sigma} - p_\lambda q_\sigma) + O(m_W^2)] \\
&\times [l_\tau l_\rho - r_\tau r_\rho + (1 - \kappa)(l.r\, g_{\tau\rho} + r_\tau q_\rho) + O(m_W^2)]
\end{aligned} \tag{4.22}$$

where the symbol $O(m_W^2)$ denotes the terms in which m_W^2 is factorized (these terms thus cannot contribute to the leading divergence in (4.22)). Further, in (4.22) we use the orthogonality of polarization vectors to the corresponding four-momenta, i.e. $k.\varepsilon(k) = 0$ etc. Thus we get finally

$$\begin{aligned}
\mathcal{M}_a^{(2)} &= -\frac{e^2}{m_W^2}\frac{1}{q^2 - m_W^2} \times \\
&\times \{(1 - \kappa)^2[(k.p)(\varepsilon(k).\varepsilon^*(p)) - (k.\varepsilon^*(p))(p.\varepsilon(k))] \\
&\times [(l.r)(\varepsilon(l).\varepsilon^*(r)) - (l.\varepsilon^*(r))(r.\varepsilon(l))] + O(m_W^2)\}
\end{aligned} \tag{4.23}$$

Now it is easy to analyse the high-energy behaviour of the considered scattering amplitude in dependence on the value of κ. First of all, it is seen that if at least one of the W's has longitudinal polarization, the potential leading divergence (quartic or cubic) vanishes for an arbitrary value of κ. This statement can be verified immediately if we replace the polarization vector $\varepsilon_L(k)$ or $\varepsilon_L(l)$ in Eq. (4.23) by the corresponding leading term k/m_W or l/m_W according to the by-now-familiar formula (H.25). The corresponding expression in square brackets is then equal to zero and thus the whole would-be leading divergence in Eq. (4.23) is suppressed. If both vector bosons $W^\pm$ have a transverse polarization, the expressions in the square brackets in Eq. (4.23) are in general non-zero and the leading term in the amplitude $\mathcal{M}_a^{(2)}$, (which in this case would be quadratically divergent) vanishes only for $\kappa = 1$. The results we have obtained may also easily be generalized to other binary processes of the considered type (in this connection see the problem 4.3). We have thus arrived at the following remarkable statement concerning tree-level diagrams of binary processes within the framework of charged vector boson electrodynamics:

Leading power divergences arising in the high-energy limit in tree-level diagrams involving both external and internal lines of vector bosons $W^\pm$ are eliminated for an arbitrary combination of the $W^\pm$ polarizations if and only if the corresponding electromagnetic interaction is of the Yang-Mills type.

Moreover, it can be shown that e.g. in the case of the considered process in Fig. 6 the resulting tree-level amplitude is asymptotically constant, i.e. it is finite in the high-energy limit for an arbitrary combination of $W^\pm$ polarizations if the vertex $WW\gamma$ is of the Yang-Mills type (i.e. the remaining non-leading divergences from diagrams (a), (b) are in such a case compensated by the diagram (c) - see the problem 4.4). Of course, the same result may be obtained also for the "Compton scattering" process $\gamma W \to \gamma W$. Thus, electromagnetic interaction of the Yang-Mills type is "optimal" in the above-specified sense (with respect to the processes considered so far).

It is important to realize that the above statement concerning the elimination of leading divergences is only valid for the tree diagrams involving *both external and internal W* lines. For tree-level diagrams involving W's in the external lines only (combined with an internal photon line) there is no general mechanism (within the framework of the electrodynamics alone) which would eliminate high-energy divergences arising from longitudinal polarizations (though, as we will see below, in some particular cases an "accidental" suppression of leading asymptotic terms may occur see also the problem 4.3). So e.g. the process $WW \to WW$ is described by the tree-level diagrams shown in Fig. 7.

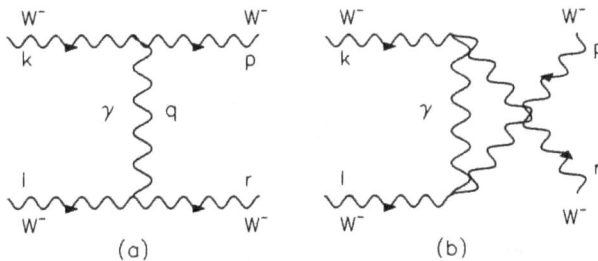

Fig. 7. Tree diagrams corresponding to the process $WW \to WW$ in the vector-boson electrodynamics.

If all four external lines correspond to longitudinally polarized W's, one may expect in general that leading asymptotic terms in both diagrams will diverge as E^4, since each external line contributes a factor of m_W^{-1} from the decomposition (H.25). If the interaction $WW\gamma$ is of the Yang-Mills type, then the anticipated quartic divergence indeed occurs; a direct calculation leads to the result

$$\mathcal{M}_a^{(YM)} + \mathcal{M}_b^{(YM)} = \frac{e^2}{4m_W^4}(t^2 + u^2 - 2s^2) + O\left(\frac{E^2}{m_W^2}\right) + O(1) \qquad (4.24)$$

where $t = (k-p)^2$, $u = (k-r)^2$. The expression for the next-to-leading quadratically divergent term $O(E^2)$ is rather complicated and we will not need it now (see however Appendix J). The following remark is in order here: If we consider the *minimal* electromagnetic interaction (i.e. $\kappa = 0$ in (4.13)) instead of the Yang-Mills $WW\gamma$ interaction, then when all the W's in diagrams in Fig. 7 have longitudinal polarizations, we get instead of (4.24)

$$\mathcal{M}_a^{(min.)} = O(1), \quad \mathcal{M}_b^{(min.)} = O(1) \qquad (4.25)$$

i.e. for $\kappa = 0$ the expected quartic divergence is completely suppressed and contributions of the relevant diagrams are - in this particular case - asymptotically constant in the high-energy limit! However, such an elimination of divergent terms only occurs when both external lines in the vertex $WW\gamma$ carry longitudinal polarizations; if e.g. transverse and longitudinal polarizations of external particles are combined in such a vertex (together with an incoming internal photon line) some divergent terms in general remain for any value of the parameter κ in (4.13) (cf. the problem 4.3). In Fig. 8 we have shown the configurations of lines entering the $WW\gamma$ vertex in corresponding diagrams, for which a divergence cancellation occurs for the Yang-Mills and the minimal electromagnetic interaction $WW\gamma$ respectively.

Salient points of the preceding discussion may be concisely summarized as follows: Electromagnetic interaction of the Yang-Mills type represents in a sense an optimal choice for the vector bosons W as it systematically eliminates leading high-energy divergences (i.e. leading powers of E for $E \to \infty$) in tree-level diagrams involving both external and internal W lines. The minimal electromagnetic interaction leads to an "accidental" suppression of divergent terms in other cases, but only for special combinations of polarizations of external W's. However, within the framework of the pure electrodynamics of charged vector bosons there is no choice of the parameter κ in (4.13) which would guarantee a cancellation of the power-like divergences in all tree-level amplitudes of binary processes.

Thus, from the point of view of high-energy behaviour of the tree diagrams, the electrodynamics of charged massive spin-1 particles (i.e. IVB's) is technically unsatisfactory in a similar way as the model of weak interactions described previously. As we have already mentioned earlier in this chapter, the quantum electrodynamics of vector bosons is non-renormalizable in higher orders of perturbation expansion. This fact is suggested by the values of effective indices for interaction vertices $WW\gamma$ and

$WW\gamma\gamma$; in both cases we obtain $\omega_v = 6$ according to the formula (G.14) in Appendix G. In the case of the Yang-Mills $WW\gamma$ interaction some types of ultraviolet divergences (coming from different diagrams) cancel [27], but even this variant of the theory has ultimately proved to be non-renormalizable [26]. The electrodynamics of IVB's thus provides another example of a connection between the "bad" high-energy behaviour of tree diagrams and non-renormalizability of perturbation expansion.

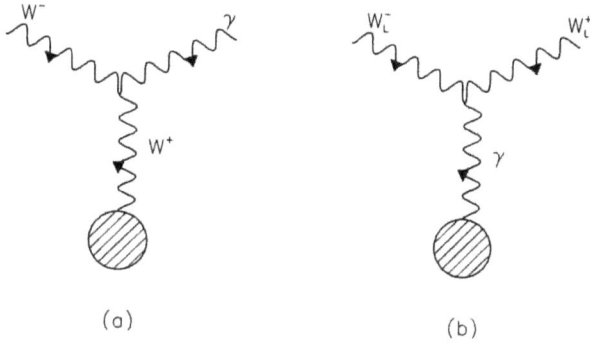

(a)   (b)

Fig. 8. If the vertex $WW\gamma$ in the configuration (a) is multiplied by the longitudinal part of the W propagator, m_W^2 is factorized (which compensates the m_W^{-2} from the propagator) for an arbitrary polarization of the external W if and only if the electromagnetic interaction is of the Yang-Mills type. In the configuration (b) the leading asymptotic term for longitudinally polarized $W^\pm$ (proportional to m_W^{-2}) vanishes only for the minimal electromagnetic interaction.

4.3 Naive Electro-Weak Unification

To close this chapter, we will now discuss some processes involving vector bosons $W^\pm$ and charged fermions. Both electromagnetic and weak interaction contribute to these processes and thus it is natural to consider a straightforward unification of weak and electromagnetic interactions described by the interaction Lagrangian

$$\mathcal{L}_{int}^{(e-w)} = \mathcal{L}_{int}^{(w)} + \mathcal{L}_{int}^{(em)} \tag{4.26}$$

where the first term in (4.26) is the weak interaction and the second term corresponds to electromagnetic interactions of charged leptons (here we will consider only the electron) and vector bosons $W^\pm$, i.e.

$$\mathcal{L}_{int}^{(em)} = -e\bar{e}\gamma_\mu e A^\mu + \mathcal{L}_{WW\gamma} + \mathcal{L}_{WW\gamma\gamma} \tag{4.27}$$

(see definitions (4.8) — (4.10)). Unless stated otherwise, we always have in mind the $WW\gamma$ interaction of the Yang-Mills type (just for comparison, we will sometimes also refer to the minimal electromagnetic interaction of W's). We will use a provisional technical term "theory of electro-weak interactions" for the model (4.26) (the hyphen indicates a superficial nature of such a facile "unification"). Binary processes in which vector bosons $W^\pm$ and charged fermions participate are essentially of two types: $\nu e \to W\gamma$ and $e^- e^+ \to W^- W^+$. Let us first consider a process of the first type, for definiteness in the configuration $\bar{\nu} e \to W^- \gamma$. Tree-level diagrams for this process corresponding to the 2nd order of perturbation expansion with respect to the interaction (4.26) are shown in Fig. 9. We will now investigate the high-energy

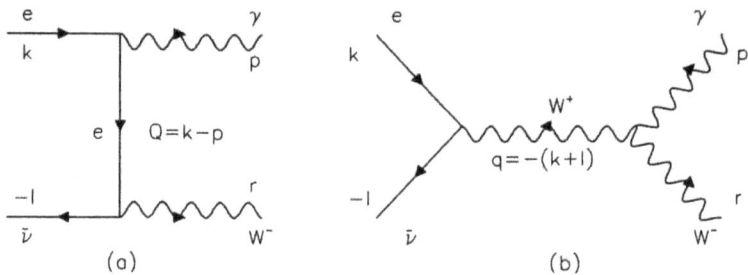

Fig. 9. *Tree diagrams of the process* $\bar{\nu} e \to W^- \gamma$.

behaviour of the corresponding scattering amplitude. First of all, from our previous results it is clear that if the final-state W^- has a transverse polarization then the contributions of both diagrams in Fig. 9 are finite in the limit $E \to \infty$. Let us further consider the case when W^- has longitudinal polarization. One may expect that the contribution of the diagram (a) contains a term linearly divergent for $E \to \infty$. As regards the diagram (b), it is not difficult to show that its part involving the factor m_W^{-2} from the corresponding propagator is finite for $E \to \infty$ (to see this, one has to realize that in this part the electron mass is also factorized — cf. (3.4)). However, the part corresponding to the diagonal term in the W propagator may yield a (linear) divergence for $E \to \infty$. Using the standard high-energy decomposition of the longitudinal polarization vector (3.20), the contribution of the diagram (a) may be written as

$$\mathcal{M}_a = \frac{eg}{2\sqrt{2}} \frac{1}{m_W} \bar{v}(l) \not{r}(1 - \gamma_5) \frac{1}{\not{Q} - m} \not{\epsilon}^*(p) u(k) + O(1) \tag{4.28}$$

With the help of tricks similar to those which in Chapter 3 have led to the relation (3.23) we get from Eq. (4.28) easily

$$\mathcal{M}_a = \frac{eg}{2\sqrt{2}} \frac{1}{m_W} \bar{v}(l) \not{\epsilon}^*(p)(1 - \gamma_5) u(k) + O(1) \tag{4.29}$$

For the contribution of the diagram (b) one may write first

$$\mathcal{M}_b = -\frac{eg}{2\sqrt{2}}\frac{1}{m_W}\bar{v}(l)\gamma_\rho(1-\gamma_5)u(k)\frac{-g^{\rho\nu}}{q^2-m_W^2}V_{\lambda\mu\nu}(p,r,q)r^\mu\varepsilon^{\lambda*}(p) + O(1) \quad (4.30)$$

where the expression $V_{\lambda\mu\nu}(p,r,q)$ is given by the formula (4.15). With the help of the 't Hooft identity (4.19), using relations $p.\varepsilon^*(p) = 0$, $p^2 = 0$ and applying the Dirac equation in the lepton matrix element, the expression (4.30) may eventually be recast as

$$\mathcal{M}_b = -\frac{eg}{2\sqrt{2}}\frac{1}{m_W}\bar{v}(l)\rlap{/}{\varepsilon}^*(p)(1-\gamma_5)u(k) + O(1) \quad (4.31)$$

Thus, it is clear from (4.29) and (4.31) that linear divergences arising in diagrams (a) and (b) cancel each other and the full tree-level amplitude is finite for $E \to \infty$, i.e.

$$\mathcal{M}_a + \mathcal{M}_b = O(1) \quad (4.32)$$

The calculation we have just performed is the first and simplest example of a divergence cancellation between tree-level diagrams of different type (the diagram (a) represents a fermion exchange in t-channel, while (b) corresponds to the s-channel exchange of vector boson). The cancellation of divergences in this case does not impose any restriction on coupling constants, as the contributions of both diagrams are proportional to $e \cdot g$. In the next chapter we will encounter many similar examples in situations where the requirement of cancellation of high-energy divergences implies nontrivial relations among coupling constants.

Let us now consider the process $e^-e^+ \to W^-W^+$. Within the framework of the theory of electro-weak interactions (4.26) it is described (in lowest order) by the tree diagrams shown in Fig. 10. The diagram (a) represents a "pure weak" and (b) "pure electromagnetic" contribution to the considered process. The worst high-energy behaviour of the corresponding amplitudes may be expected when both vector bosons $W^\pm$ have longitudinal polarizations; one may then guess, in the same way as in the preceding examples, that both diagrams in Fig. 10 may contain quadratically divergent terms for $E \to \infty$: (However, let us recall that if the $WW\gamma$ vertex corresponded to the minimal electromagnetic interaction, quadratic divergence in the diagram (b) would vanish — see Fig. 8.) When only one of W's has longitudinal polarization, both diagrams (a), (b) yield linear divergences for $E \to \infty$ (for the Yang-Mills $WW\gamma$ vertex as well as for the minimal electromagnetic interaction). We will now examine in more detail the case when both vector bosons $W^\pm$ have longitudinal polarizations. For the contribution of the diagram (a) we get, using Eq. (3.20) and after usual manipulations

$$\mathcal{M}_a = -\frac{g^2}{4m_W^2}\bar{v}(l)\rlap{/}{p}(1-\gamma_5)u(k) + O\left(\frac{m}{m_W^2}E\right) + O(1) \quad (4.33)$$

As we have indicated in Eq. (4.33), the amplitude $\mathcal{M}_a$ contains the leading quadratic divergence as well as a next-to-leading (linear) divergence for $E \to \infty$ (cf. in this

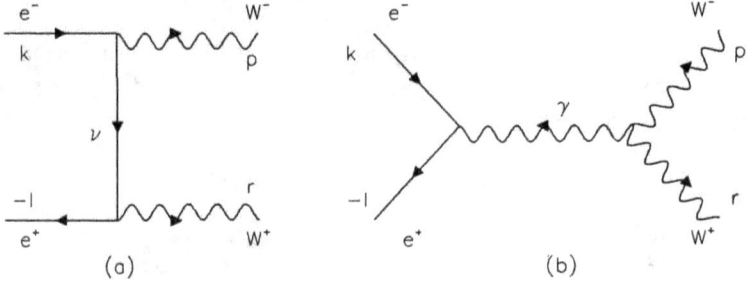

Fig. 10. Tree diagrams corresponding to the process $e^- e^+ \to W^- W^+$.

context the remarks concerning the relation (3.25) in previous chapter). Derivation of an explicit form of the linearly divergent term is left to the reader as an easy exercise (see also Appendix J, the formula (J.1)). For the contribution of the diagram (b) (by means of manipulations similar to those which have led from (4.30) to (4.31)) one gets the result

$$\mathcal{M}_b = \frac{e^2}{m_W^2} \bar{v}(l) \not{p} u(k) + O(1) \tag{4.34}$$

If we now compare (4.33) and (4.34) it is clear that one cannot accomplish a mutual cancellation of quadratic divergences in $\mathcal{M}_a$ and $\mathcal{M}_b$ by any clever choice of the relative magnitude of the coupling constants e and g since the corresponding expression in (4.33) contains $1 - \gamma_5$ but (4.34) does not; this means that quadratically divergent terms in $\mathcal{M}_a$ and $\mathcal{M}_b$ depend differently on lepton polarizations. Besides that, in the expression (4.34) there is no linearly divergent term of the type $O(mE/m_W^2)$, in contrast with (4.33); of course, this is a consequence of the conservation of lepton electromagnetic current in the corresponding vertex of the diagram (b). We thus see that the full tree-level amplitude of the process $e^- e^+ \to W^- W^+$ contains (if at least one of the W's has longitudinal polarization) terms diverging as a positive power of energy for $E \to \infty$.

We will now summarize the main results concerning the high-energy behaviour of tree-level amplitudes of binary processes, that we have obtained in this and the preceding chapter. The naive theory of weak interactions with charged IVB and the electrodynamics of IVB have similar problems with power-like growth of tree-level amplitudes for $E \to \infty$. Trivial unification of weak and electromagnetic interactions in the Lagrangian (4.26) does not solve these problems. For the process $e^- e^+ \to W^- W^+$, a cancellation between leading divergences coming from the weak and electromagnetic contributions respectively is not possible, because the weak interaction violates parity (via $V - A$ currents) while the electromagnetic interaction is parity conserving. In other cases one has only a weak contribution (e.g. for $\nu\bar{\nu} \to W^-W^+$) or an electromagnetic one (e.g. for $WW \to WW$) and the terms divergent for $E \to \infty$ cannot be eliminated trivially. It is obvious that the power-like high-energy growth of the above-mentioned tree amplitudes cannot be suppressed without introducing new particles and new interactions which represent the "missing links" of the naive model of electro-weak interactions. Keeping in mind the remarks we have made concerning the process $e^-e^+ \to W^-W^+$ one may guess that the necessary new interactions should in a sense "interpolate" between the original weak and electromagnetic interactions in (4.26). A detailed construction of the "missing links" of the unified theory of weak and electromagnetic interactions is the subject of the next chapter.

Problems

4.1. Prove the 't Hooft identity (4.19).

4.2. Prove the relation (4.24).

4.3. Prove the statement in the text of Fig. 8.

4.4. Show that full tree-level amplitude of the process $W^-W^+ \to \gamma\gamma$ is finite in the high-energy limit for an arbitrary combination of $W^\pm$ polarizations if the $WW\gamma$ interaction is of the Yang-Mills type. Can there be a cancellation of non-leading divergences in diagrams (a), (b), (c) in Fig. 6 when the $WW\gamma$ vertex is described by the expression (4.13) with a parameter $\kappa \neq 1$?

4.5. Derive (4.29) and (4.31).

4.6. Calculate the leading term in the cross section of the process $\bar{\nu}e \to W^-\gamma$ for $E \to \infty$ in the approximation of tree diagrams in Fig. 9 (for unpolarized particles).

4.7. Derive (4.33) and (4.34).

4.8. Calculate the leading term in the cross section $e^-e^+ \to W^-W^+$ for $E \to \infty$ in the approximation of tree diagrams in Fig. 10 (for unpolarized particles). What are the separate contributions of the weak and electromagnetic interaction ? (see also problem 3.6). How are the corresponding results changed if we consider the minimal electromagnetic interaction instead of the Yang-Mills vertex $WW\gamma$?

Chapter 5

TREE UNITARITY AND ELECTROWEAK INTERACTIONS

5.1 Criterion for Perturbative Renormalizability

We have shown in preceding chapters that the naive hypothesis on the existence of charged intermediate vector boson of weak interactions leads eventually - despite partial successes - to similar difficulties as the original Fermi-type theory. Moreover, the introduction of an electromagnetic interaction of IVB modifies substantially the properties of quantum electrodynamics: Contrary to the familiar spinor QED, electrodynamics of charged massive spin-1 particles is non-renormalizable (and, at the same time, some tree-level amplitudes display a "bad" high-energy behaviour). In this chapter we will demonstrate that a non-trivial unification of weak and electromagnetic interactions (which necessitates postulating extra particles and a host of new terms in the interaction Lagrangian) is able to cure simultaneously the difficulties of the old provisional models of $W^\pm$ interactions, i.e. of both the electrodynamics and weak interaction theory.

Let us now specify our goal more precisely. We wish to construct a physically realistic theory of weak and electromagnetic interactions (i.e. such that it correctly describes experimental data e.g. for muon decay, Compton scattering etc.) and we require that the model would be renormalizable within the framework of perturbation expansion. The following remark is in order here: The requirement of perturbative renormalizability is in fact of a technical nature and it is not clear at present whether it is indeed physically relevant in its full extent. Nevertheless, this technical requirement proved to be an extremely valuable heuristic principle which has led to many non-trivial physical predictions (some of which have already been verified experimentally).

However, a direct search for a renormalizable model of weak and electromagnetic interactions would be a tremendous task: It would amount to a systematic analysis of ultraviolet divergences in Feynman diagrams involving at least one closed loop and to finding conditions of a cancellation of non-renormalizable divergences descending from different diagrams. From the technical point of view, it is much easier to employ a connection between perturbative renormalizability and the high-energy behaviour of *tree-level* diagrams which has been observed in the discussion of the models described in preceding chapters. We will now formulate the relevant necessary condition for perturbative renormalizability in detail (cf. the end of Chapter 2) and at the same time we will introduce a terminology commonly used in the literature (see [11] - [14]).

The experience gained from various quantum field theory models suggests that a necessary condition for the renormalizability of perturbation expansion is "asymptotic softness" of tree-level scattering amplitudes [14] or, in other words, "tree unitarity" [11 - 14]: Such a condition means that an arbitrary n-point tree-level amplitude $\mathcal{M}^{(n)}_{tree}$ (i.e. the amplitude of a process $1 + 2 \to 3 + 4 + \ldots + n$ in the approximation of tree diagrams) behaves (for fixed non-zero scattering angles) in the limit $E \to \infty$ at most like

$$\mathcal{M}^{(n)}_{tree} = O(E^{4-n}) \tag{5.1}$$

(cf. relation (C.3) for the dimension of $\mathcal{M}^{(n)}$). In particular, for binary processes the condition (5.1) means that the corresponding (dimensionless) amplitude is asymptotically flat at high energies, i.e.

$$\mathcal{M}^{(4)}_{tree} = O(1), \tag{5.2}$$

for the amplitude of a process $1 + 2 \to 3 + 4 + 5$ in the limit $E \to \infty$ one should have

$$\mathcal{M}^{(5)}_{tree} = O\left(\frac{1}{E}\right) \tag{5.3}$$

etc. In the subsequent discussion the condition (5.2) (which we have already mentioned in preceding chapters) will be applied in a detailed manner to many particular processes and finally we will also mention an application of the condition (5.3).

As regards the high-energy behaviour of the full amplitude $\mathcal{M}^{(n)}$ in a renormalizable theory (to an arbitrary fixed order of perturbation theory) its power-law character expressed by (5.1) is modified in higher orders at most logarithmically (cf. [13]), i.e.

$$\mathcal{M}^{(n)}|_{E\to\infty} = O(E^{4-n} \ln^k E) \tag{5.4}$$

where $k \geq 0$.

The term "tree unitarity" of course does not mean that e.g. a four-point scattering amplitude satisfying the condition (5.2) also fulfills exactly the unitarity condition (see (E.12) or (E.15)); one has to keep in mind that in a fixed order of perturbation expansion, unitarity of S-matrix is in general always violated. The technical term we are using refers to the fact that fulfilling the condition (5.2) for $E \to \infty$ implies, in a sense, a "minimal" unitarity violation in the tree approximation: In such a case, partial-wave amplitudes in the Jacob-Wick expansion grow at worst logarithmically for $E \to \infty$ (cf. the examples in Chapter 3). An equivalent term "asymptotic softness of tree-level amplitudes" [14] is more straightforward and thus perhaps more instructive, but it is not commonly used in the literature.

The tree unitarity (5.1) thus represents a definite criterion for perturbative renormalizability which is particularly valuable in the case of interactions of charged massive vector bosons. This criterion seems to be generally accepted but one has to stress that it is not completely rigorous. It is based on the observation that in all known renormalizable models of quantum field theory the condition (5.1) is satisfied and, moreover, there is a plausible intuitive argument in its favour. We will now give this argument, which is obviously superficial but still rather instructive (cf. [13] and also [18]).

Higher-order diagrams (i.e. those involving at least one closed loop) are obtained, in a sense, by means of an iteration of tree diagrams: The imaginary part of a one-loop graph may be expressed, roughly speaking, in terms of an appropriate tree-level amplitude squared, from tree-level and one-loop graphs one may get imaginary part of a two-loop diagram etc. Such an iteration procedure of course corresponds to unitarity conditions for the S-matrix within the framework of perturbation expansion (see e.g. [16], [20], [21]) and one example of this kind is depicted schematically in Fig. 11. Thus, if the tree-level amplitude of some binary process behaved for $E \to \infty$ as E^δ, where $\delta > 0$, then the imaginary part of a one-loop amplitude (corresponding in general to a different, appropriately chosen process cf. Fig. 11) would behave like $E^{2\delta}$, i.e. it would grow faster than the tree approximation in the limit $E \to \infty$. From the imaginary part of a diagram one may calculate the full amplitude via a dispersion relation (see e.g. [16], [20], [21]); in doing this, one has to perform appropriate subtractions in order to suppress (ultraviolet) divergences. The essential point is that - as we have already observed - the power-like growth of one-loop amplitudes is in general "worse" than that encountered on the tree level. In further iterations (i.e. for more complicated diagrams) the power behaviour of the corresponding imaginary parts gets worse, which necessitates introducing more subtractions in dispersion relations; this in turn corresponds to an infinite number of renormalization counterterms, i.e. to the perturbation expansion which is not renormalizable in the usual sense. On the other hand, if the tree-level amplitudes of binary processes satisfy the condition (5.2), the imaginary parts of one-loop diagrams in general behave for $E \to \infty$ in the same way as the tree-level amplitudes and there is *a priori* no manifest reason to expect that the character of the power behaviour would be substantially changed at higher orders. In fact, however, it may happen that (as a consequence of the integration in a dispersion relation) the high-energy asymptotics of the real part of a one-loop amplitude is different from that of the imaginary part; in such a way one may encounter a situation in which the condition (5.2) is fulfilled but some one-loop amplitudes grow as a positive power of energy for $E \to \infty$ (i.e. the relation (5.4) is then violated). To be more specific, the envisaged situation is known to occur owing to the presence of the famous Adler-Bell-Jackiw (ABJ) triangle anomaly [40]; this remarkable phenomenon will be discussed in more detail in Section 5.6 (see also [17]).

The intuitive arguments which we have given thus indeed indicate that the tree

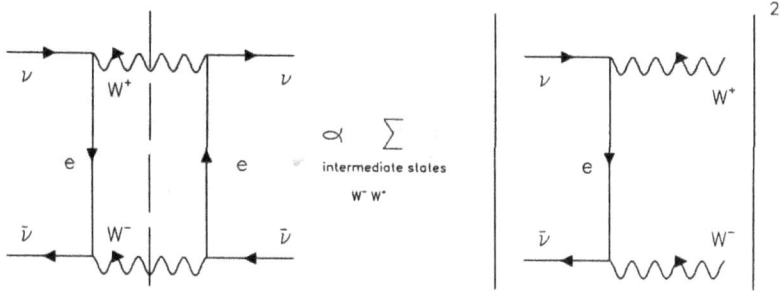

Fig. 11. A connection between the imaginary part of a one-loop diagram for the process $\nu\bar{\nu} \to \nu\bar{\nu}$ and the tree-level amplitude of the process $\nu\bar{\nu} \to W^-W^+$ in the naive model of weak interactions with charged IVB.

unitarity expressed by the relation (5.1) is a necessary condition for perturbative renormalizability; however, one may find explicit examples showing that it is not a sufficient condition.

Finally let us remark that the condition of tree unitarity may also be pragmatically understood (apart from its deep connection with renormalizability) as follows: If (5.1) holds, then the tree approximation is not in an obvious conflict with the general requirement of S-matrix unitarity in a "maximum" energy range (which corresponds to at worst logarithmic growth of partial-wave amplitudes), i.e. the tree approximation is then applicable for all "terrestrial" energies.

The exposition of the following paragraphs is conceptually very close to Refs. [14], [18] and [39] (the influence of the classic lecture notes [18] was particularly stimulating) but in fact it is independent of these sources.

5.2 Mechanisms of Divergence Cancellations

Let us now consider again the process $e^-e^+ \to W^-W^+$ when both vector bosons $W^{\pm}$ have longitudinal polarization. If one wants to eliminate the leading (quadratic) divergences arising in the limit $E \to \infty$ in the weak and electromagnetic contributions to the corresponding tree-level amplitude (see Fig. 10 and the relations (4.33), (4.34)), one obviously has to postulate the existence of a new particle and corresponding interactions. We will *a priori* restrict ourselves to particles with a lowest possible spin, i.e. $0, \frac{1}{2}$ or 1 and we will consider only the interaction terms satisfying the condition (see Appendix G)

$$\dim \mathcal{L}_{int} \leq 4 \tag{5.5}$$

so as not to introduce any other potential source of a non-renormalizable behaviour of Feynman diagrams at higher orders of the perturbation expansion; in other words, we will only solve the problems due to the presence of charged massive vector bosons.

First let us consider postulating a (neutral) spin-0 particle as an attempt to cure the quadratically divergent terms in the expressions (4.33) and (4.34). We will denote the corresponding (real) field as η. In order to be able to draw a Feynman diagram involving an exchange of the spinless particle contributing to the amplitude of the process (see Fig. 12), one has to introduce interaction terms of the type $WW\eta$ and $ee\eta$. It is not difficult to realize that the only possible choice preserving the condition (5.5) (as well as the Lorentz invariance) is then

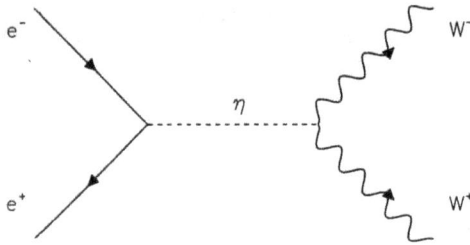

Fig. 12. The lowest-order Feynman diagram for $e^-e^+ \to W^-W^+$ involving the exchange of a neutral spin-0 particle.

$$\mathcal{L}_{WW\eta} = g_{WW\eta} W_\mu^- W^{+\mu} \eta \tag{5.6}$$

and

$$\mathcal{L}_{ee\eta} = g_{ee\eta} \bar{e} \Gamma e \eta \tag{5.7}$$

where Γ in the Yukawa-type interaction (5.7) is in general a combination of the γ_5 and the unit matrix and $g_{WW\eta}$, $g_{ee\eta}$ are the corresponding coupling constants. It is important to notice that the coupling constant $g_{WW\eta}$ is not dimensionless (contrary to the $g_{ee\eta}$); one obviously has (cf. Appendix G)

$$[g_{WW\eta}] = M \tag{5.8}$$

in units of an arbitrary mass. As a consequence of this, the diagram in Fig. 12 can diverge at most linearly for $E \to \infty$ in the case of longitudinally polarized vector bosons, since the coupling constant $g_{WW\eta}$ compensates one of the factors of m_W^{-1} from $W^\pm$ polarizations; the contribution of Fig. 12 thus behaves at worst like $O(g_{WW\eta} E / m_W^2)$ in the limit $E \to \infty$. An exchange of a spin-0 particle is therefore not sufficient for the desirable compensation of the quadratic divergences in (4.33) and (4.34). (However,

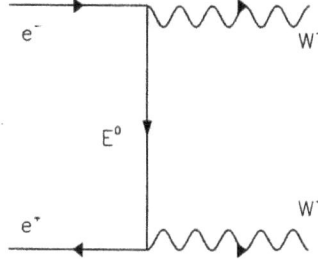

Fig. 13. The exchange of a hypothetical neutral heavy lepton in the process $e^- e^+ \rightarrow W^- W^+$.

such an exchange is able to suppress linear divergences which may eventually occur and it will play an important role later.)

As another possibility, let us now consider instead of Fig. 12 an exchange of a neutral spin-$\frac{1}{2}$ particle, i.e. of a hypothetical "heavy lepton" E^0. The corresponding diagram is shown in Fig. 13 (cf. the analogous Fig. 10(a)). The most general interaction term producing the diagram in Fig. 13 is given by

$$\mathcal{L}_{int}^{(E^0)} = (f_L \bar{E}_L^0 \gamma^\mu e_L + f_R \bar{E}_R^0 \gamma^\mu e_R)W_\mu^+ + \text{h.c.} \tag{5.9}$$

where the index L or R denotes the left-handed or right-handed component of the corresponding fermion field respectively and h.c. means the hermitian conjugate. In contrast to the preceding case, the contribution of Fig. 13 for longitudinally polarized $W^\pm$ does contain terms quadratically divergent in the high-energy limit. The requirement of a compensation of quadratic divergences in the expressions (4.33) and (4.34) then yields the following conditions for the coupling constants f_L, f_R (see also [18]):

$$\begin{aligned} f_L^2 &= e^2 - \frac{1}{2}g^2 \\ f_R^2 &= e^2 \end{aligned} \tag{5.10}$$

The first relation in (5.10) thus leads to a constraint for relative strength of weak and electromagnetic interactions, namely

$$g \leq e\sqrt{2} \tag{5.11}$$

One can see from (5.10) that the interaction of the heavy lepton E^0 "interpolates" between the original weak and electromagnetic interaction (as we have anticipated in the preceding chapter) and in this sense a unification of the two forces is indeed realized. The condition (5.11) guarantees the existence of a real solution of eq. (5.10) and thus it is natural to call it a "unification condition". An interesting consequence of the inequality (5.11) and of the general relation $m_W^2 = g^2(4G_F\sqrt{2})^{-1}$ (see (3.7)) is an *upper* bound for the $W^\pm$ mass:

$$m_W \leq \left(\frac{\pi\alpha\sqrt{2}}{G_F}\right)^{\frac{1}{2}} \doteq 53\text{GeV} \qquad (5.12)$$

In this way we could proceed in eliminating systematically the diverging terms for all relevant scattering processes. It turns out that the alternative of heavy leptons leads indeed to the desired goal (without introducing new massive vector bosons); within the indicated scheme one would thus arrive at a "minimal" renormalizable model of this type which was originally invented by Georgi and Glashow [41] and formulated as the corresponding non-abelian gauge theory with Higgs mechanism. However, such a model is — as we shall see later — in striking disagreement with experimental facts. The scenario of heavy leptons, though theoretically plausible (and even appealing) thus obviously does not correspond (at least in its simplest version) to the real world. For this reason we will not consider this scheme further, although from a technical point of view it represents a remarkable and instructive example of a renormalizable model of the unification of weak and electromagnetic interactions (the interested reader may find further details in the original paper [41] and also in [15] and [18]).

Finally, we shall examine the last remaining possibility, i.e. where the "compensation" diagram for the considered process $e^-e^+ \to W^-W^+$ corresponds to an exchange of a neutral spin-1 particle with non-zero mass (the exchange of a massless particle would lead to a new type of long-range force which is not observed in nature); this neutral vector boson will be denoted as Z. The corresponding diagram is depicted in Fig. 14.

Let us first estimate the asymptotic behaviour of the contribution of Fig. 14 for $E \to \infty$ when both vector bosons $W^\pm$ have longitudinal polarizations. The worst divergence might obviously arise from the term involving the longitudinal part of the Z propagator, i.e. from the part proportional to $q^\mu q^\nu$. In the limit $E \to \infty$ this term behaves in general like $O(m_Z^{-2}m_W^{-2}mE^3)$ because one of the factors q^μ, q^ν acts on the lepton vertex and an application of the Dirac equation leads to a factorization of the electron mass (cf. (3.4)). Thus, in contrast to the quadratically diverging expressions (4.33), (4.34), the contribution of Fig. 14 may in general contain a *cubic* divergence for $E \to \infty$. We have already encountered a similar problem in the framework of the electrodynamics of charged IVB (cf. the discussion around Fig. 6 in the preceding chapter). The leading divergent term in the contribution of Fig. 14 (and in all the other diagrams which one must consider as a consequence of introducing the interaction WWZ) can be eliminated by means of an appropriate choice of the

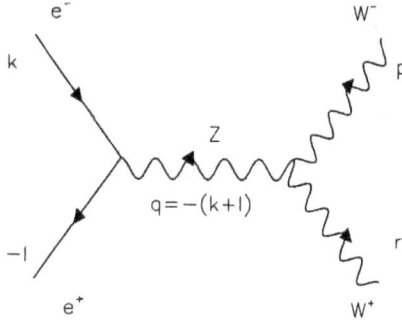

Fig. 14. The exchange of a neutral vector boson in the process $e^-e^+ \to W^-W^+$.

interaction vertex WWZ in complete analogy with the case of the electromagnetic interaction $WW\gamma$. Namely, the following statement is valid:

Leading power-like divergences arising in the high-energy limit in tree-level diagrams involving interaction vertices of the type WWZ vanish for an arbitrary combination of polarizations of external $W^\pm$ and Z if and only if the trilinear vector-boson interaction WWZ is of the Yang-Mills type, i.e. if the corresponding interaction Lagrangian has the form

$$\mathcal{L}_{WWZ} = -ig_{WWZ}(Z^\mu W^{-\nu}\overleftrightarrow{\partial}_\mu W^+_\nu + W^{-\mu}W^{+\nu}\overleftrightarrow{\partial}_\mu Z_\nu + W^{+\mu}Z^\nu\overleftrightarrow{\partial}_\mu W^-_\nu) \quad (5.13)$$

where g_{WWZ} is a (real) coupling constant.

A proof of this statement is briefly sketched in Appendix I. However, for completeness let us add that e.g. for the diagram in Fig. 14 the would-be leading divergence is in fact suppressed not only for (5.13) but also for a wider class of $W\hat{W}Z$ interactions. (As we have seen in the preceding chapter, a similar situation occurs in some particular cases also for the electromagnetic interaction $WW\gamma$.) The essential feature of the WWZ interaction of the Yang-Mills type is that this option eliminates potential leading divergences (which could not be compensated by another diagram) *in all cases.* In what follows we shall therefore consider only the WWZ interaction (5.13).

The interaction Lagrangian (5.13), like the electrodynamics of vector bosons $W^\pm$, leads to the Feynman rule for the WWZ vertex in Fig. 15

$$\mathcal{V}_{\lambda\mu\nu}(k,p,q) = g_{WWZ}V_{\lambda\mu\nu}(k,p,q) \quad (5.14)$$

where the expression $V_{\lambda\mu\nu}(k,p,q)$ is defined by the relation (4.15).

As we have already stated, the would-be cubic divergence in the contribution of Fig. 14 can be made to vanish, so now we may consider a possible compensation of the quadratically divergent terms in (4.33) and (4.34). In the next section we will formally define the corresponding interactions of the neutral vector boson Z with leptons and we will investigate in detail the conditions for eliminating the leading power-like divergences in the expressions (4.33), (4.34) and also in tree-level diagrams for other processes. As indicated in preceding chapters, a systematic elimination of the terms violating the tree-unitarity condition (5.1) will ultimately lead to recovering the standard GWS model [5, 6, 7] of electroweak interactions; introducing a neutral IVB is an important step in this direction.

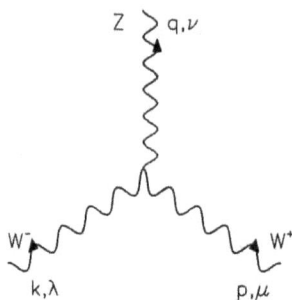

Fig. 15. Vertex corresponding to the trilinear interaction WWZ.

5.3   Weak Neutral Currents

Before a detailed discussion of the process $e^-e^+ \rightarrow W^-W^+$, we will come back to a simpler case mentioned in Chapter 3, namely to the process $\nu\bar{\nu} \rightarrow W^-W^+$. Let us consider again longitudinally polarized vector bosons $W^\pm$. We will attempt to compensate the quadratically divergent term (3.24) in the expression (3.25) for Fig. 3 by means of a diagram involving an exchange of the neutral massive vector boson Z in analogy with Fig. 14. Both relevant tree diagrams of the process $\nu\bar{\nu} \rightarrow W^-W^+$ are shown in Fig. 16 (for convenience we have also reproduced here Fig. 3). The diagram (b) in Fig. 16 corresponds to a new interaction (in addition to (5.13)) of the type

$$\mathcal{L}_{\nu\nu Z} = g_{\nu\nu Z}\, \bar{\nu}_L \gamma^\mu \nu_L\, Z_\mu \tag{5.15}$$

where $g_{\nu\nu Z}$ is the corresponding coupling constant (we still assume, for simplicity, that the neutrino is massless and therefore only the left-handed component of the corresponding field is introduced). Using (5.14), (H.25) and other standard rules one may write for the contribution of Fig. 16

$$
i\mathcal{M}_b = i^3 \frac{1}{2} g_{\nu\nu Z} \, g_{WWZ} \, \bar{v}(l)\gamma_\rho(1-\gamma_5)u(k) \times
$$
$$
\times \frac{-g^{\rho\nu}+m_Z^{-2}q^\rho q^\nu}{q^2-m_Z^2} V_{\nu\mu\lambda}(q,r,p)\frac{p^\lambda}{m_W}\frac{r^\mu}{m_W}
$$
$$
+ \; O(1) \tag{5.16}
$$

The longitudinal term from the Z propagator (i.e. the part proportional to $m_Z^{-2}q^\rho q^\nu$) does not contribute at all, irrespective of the form of the WWZ interaction (this is an automatic consequence of the assumption $m_\nu = 0$ and of the Dirac equation). Using further the 't Hooft identity (4.19), the relation (5.16) may be easily recast as

$$
\mathcal{M}_b = \frac{1}{2m_W^2} g_{\nu\nu Z} \, g_{WWZ} \, \bar{v}(l)\slashed{p}(1-\gamma_5)u(k) + O(1) \tag{5.17}
$$

A corresponding relation for the contribution of the diagram (a) in Fig. 16 has been derived in Chapter 3 (see (3.24) and (3.25)):

$$
\mathcal{M}_a = -\frac{g^2}{4m_W^2}\bar{v}(l)\slashed{p}(1-\gamma_5)u(k) + O(1) \tag{5.18}
$$

Comparing the expressions (5.17) and (5.18) one immediately gets a condition for the compensation of power-like (quadratic) high-energy divergences in the tree-level amplitude of the process $\nu\bar{\nu} \to W_L^- W_L^+$ in the limit $E \to \infty$:

$$
-\frac{1}{2}g^2 + g_{\nu\nu Z} \, g_{WWZ} = 0 \tag{5.19}
$$

It is not difficult to verify that the condition (5.19) guarantees a compensation of power-like divergences in the amplitude of the considered process for any combination of $W^\pm$ polarizations (i.e. including the case when one of the final-state W's is polarized longitudinally and the other transversely).

We will now examine in detail the tree-level amplitude for $e^-e^+ \to W^-W^+$. For convenience, all diagrams considered up to now (see Fig. 10 and Fig. 14) are reproduced in Fig. 17. The "compensation diagram" in Fig. 17(c) corresponds to a new interaction (in addition to (5.13)) of the type eeZ. For obvious reasons, we will parametrize the corresponding interaction Lagrangian by means of two coupling constants which we denote for brevity as g_L and g_R:

$$
\mathcal{L}_{eeZ} = (g_L \bar{e}_L \gamma^\mu e_L + g_R \bar{e}_R \gamma^\mu e_R)Z_\mu \tag{5.20}
$$

As we have stated, asymptotic behaviour of the contributions of Fig. 17(a), (b) in the limit $E \to \infty$ can be expressed by means of the formulae (see (4.33) and (4.34))

$$
\mathcal{M}_a = -\frac{g^2}{4m_W^2}\bar{v}(l)\slashed{p}(1-\gamma_5)u(k) + O(\frac{m}{m_W^2}E) + O(1) \tag{5.21}
$$

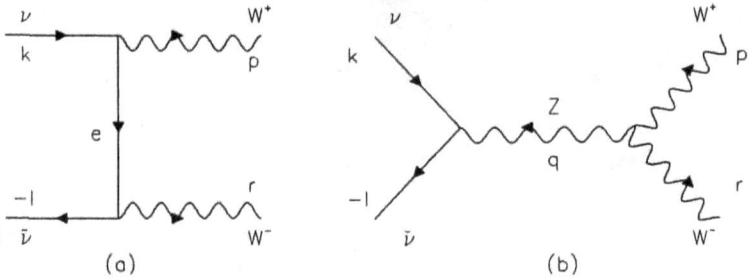

Fig. 16. (a) The diagram for $\nu\bar{\nu} \to W^- W^+$ corresponding to naive weak interaction theory with charged IVB. (b) The "compensation diagram" involving an exchange of the neutral IVB.

$$\mathcal{M}_b = \frac{e^2}{m_W^2}\bar{v}(l)\not{p}u(k) + O(1) \tag{5.22}$$

Using standard procedures (see Appendix J), from (5.20) one easily obtains the leading (quadratically divergent) asymptotic terms in the contribution of the diagram (c):

$$\begin{aligned}
\mathcal{M}_c &= -\frac{1}{2m_W^2}\, g_{WWZ}\, g_L \bar{v}(l)\not{p}(1-\gamma_5)u(k) \\
&\quad - \frac{1}{2m_W^2}\, g_{WWZ}\, g_R \bar{v}(l)\not{p}(1+\gamma_5)u(k) \\
&\quad + O\left(\frac{m}{m_W^2}E\right) + O(1)
\end{aligned} \tag{5.23}$$

Explicit expressions for the next-to-leading (i.e. linearly divergent) terms contained in Eq. (5.21) and Eq. (5.23) are given in Appendix J and we will deal with them later. From (5.21) – (5.23) one immediately gets conditions for the compensation of leading divergences for $E \to \infty$:

$$-\frac{1}{2}g^2 + e^2 - g_L\, g_{WWZ} = 0 \tag{5.24}$$

$$e^2 - g_R\, g_{WWZ} = 0 \tag{5.25}$$

(Fulfilling these relations means that the would-be quadratic divergences vanish for any combination of polarizations of the initial-state e^- and e^+.)

The relations (5.19), (5.24) and (5.25) represent three equations for the four unknown coupling constants g_{WWZ}, $g_{\nu\nu Z}$, g_L and g_R if the e and g are assumed to be known (these are the parameters of the original naive theory of electro-weak interactions). However, now one can also consider the process $\bar{\nu}e \to W^- Z$; in the 2nd order

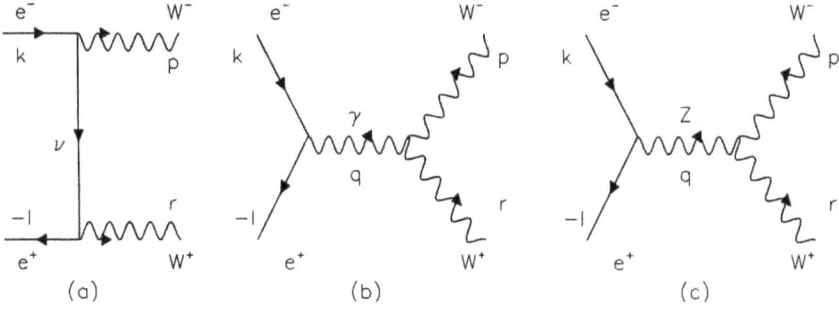

Fig. 17. The process $e^-e^+ \rightarrow W^-W^+$. (a) The contribution of weak charged-current interaction. (b) Electromagnetic contribution. (c) Exchange of the neutral IVB.

of perturbation expansion (with respect to the interaction terms introduced so far) it is described by the tree diagrams depicted in Fig. 18.

For contributions of the diagrams in Fig. 18 one gets (we give here explicitly only the leading terms, quadratically divergent for $E \rightarrow \infty$; for the subleading (linear) divergences see Appendix J).

$$\mathcal{M}_a = -\frac{gg_L}{2\sqrt{2}} \frac{1}{m_W m_Z} \bar{v}(l)\not{p}(1 - \gamma_5)u(k) + O(E) + O(1) \tag{5.26}$$

$$\mathcal{M}_b = \frac{gg_{\nu\nu Z}}{2\sqrt{2}} \frac{1}{m_W m_Z} \bar{v}(l)\not{p}(1 - \gamma_5)u(k) + O(E) + O(1) \tag{5.27}$$

$$\mathcal{M}_c = -\frac{gg_{WWZ}}{2\sqrt{2}} \frac{1}{m_W m_Z} \bar{v}(l)\not{p}(1 - \gamma_5)u(k) + O(E) + O(1) \tag{5.28}$$

The requirement of a cancellation of quadratic divergences in the sum of the expressions (5.26) – (5.28) immediately gives the condition

$$- g_L + g_{\nu\nu Z} - g_{WWZ} = 0 \tag{5.29}$$

As regards the next-to-leading (linear) divergences, the results given in Appendix J show clearly that for the process $e^-e^+ \rightarrow W_L^- W_L^+$ these cannot be eliminated by any particular choice of the relevant coupling constants. Indeed, the corresponding amplitude contains terms proportional to $\bar{v}(l)u(k)$ and $\bar{v}(l)\gamma_5 u(k)$ (see (J.1) and (J.5)), which for obvious reasons should be eliminated separately. A term of the first type (contained only in the contribution of Fig. 17(a) - see (5.1)) has an overall coefficient $-g^2 m(4m_W^2)^{-1}$ which of course cannot be zero. Thus it is seen that in this case it

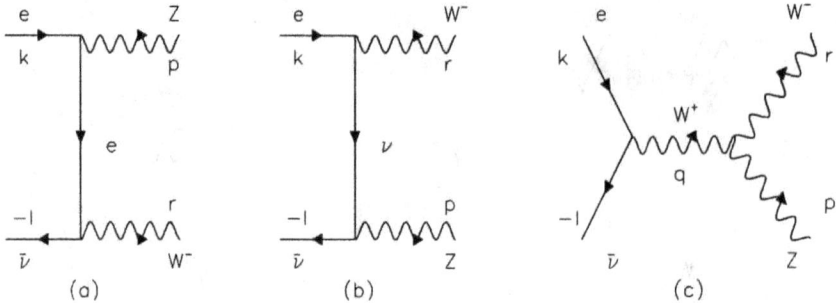

Fig. 18. The diagrams of the process $\bar\nu e \to W^- Z$.

will be necessary to introduce an additional compensation diagram involving an exchange of a new (neutral) particle to tame such residual next-to-leading divergences; as we have remarked earlier in this chapter, a spin-0 particle is sufficient for such a purpose (cf. the discussion around Fig. 12). We will return to this important problem in Section 5.5. For the process $\bar\nu e \to W_L^- Z_L$, all the linearly divergent terms in the corresponding amplitude are proportional to the expression $\bar v(l)(1 + \gamma_5)u(k)$ (see (J.6), (J.7) and (J.13)) and one may try to eliminate them by means of an appropriate choice of the ratio m_W^2/m_Z^2 (see (J.13)) as a function of coupling constants. Such a compensation would be highly desirable since our aim is to construct a "minimal" model of weak and electromagnetic interactions satisfying the condition of tree unitarity. We have already observed in the previous example that one cannot avoid introducing a new neutral spin-0 particle; if the linear divergence in the amplitude of $\bar\nu e \to W_L^- Z_L$ did not vanish owing to a suitable relation among coupling constants and masses, it would be necessary to introduce an extra spin-0 particle (which would have to be charged).

At this point one could also naturally ask what is the situation in other similar processes of the considered type, in particular e.g. $e^- e^+ \to Z_L Z_L$ or $e^- e^+ \to Z_L \gamma$. We will discuss these problems in more detail in Section 5.5; here let us only remark that the divergences arising in the corresponding tree-level amplitudes are at most linear (see e.g. (5.78)) and thus it is not necessary to introduce new direct interactions of three vector boson fields (this favourable circumstance is of course closely related to the fact that Z and γ are *neutral* particles).

For convenience, let us now summarize the equations for coupling constants of the interactions WWZ, eeZ and $\nu\nu Z$, which follow from the requirement of cancellations of the leading power-like (quadratic) divergences in the limit $E \to \infty$ in the tree diagrams of processes $\nu\bar\nu \to W_L^- W_L^+$, $e^- e^+ \to W_L^- W_L^+$ and $\bar\nu e \to W_L^- Z_L$. We have obtained four equations for the four unknowns g_{WWZ}, $g_{\nu\nu Z}$, g_L and g_R (see (5.19), (5.24), (5.25) and (5.29)):

$$-\frac{1}{2}g^2 + g_{\nu\nu Z}\, g_{WWZ} = 0$$

$$-\frac{1}{2}g^2 + e^2 - g_L\, g_{WWZ} = 0$$

$$e^2 - g_R\, g_{WWZ} = 0$$

$$-g_L + g_{\nu\nu Z} - g_{WWZ} = 0 \tag{5.30}$$

Moreover, the condition of a supposed compensation of linearly divergent terms in the amplitude of the process $\bar{\nu}e \to W_L^- Z_L$ (see (J.6), (J.7), (J.13)) is

$$g_R - g_{\nu\nu Z} + g_{WWZ}\left(1 - \frac{m_Z^2}{2m_W^2}\right) = 0 \tag{5.31}$$

First we will deal with solving the system of equations (5.30). From the first, the second and the fourth of them one can obtain easily

$$g^2 - e^2 = g_{WWZ}^2 \tag{5.32}$$

An important constraint follows immediately from Eq. (5.32), namely (cf. (5.11))

$$e < g \tag{5.33}$$

(let us emphasize that the strict inequality must hold, since for $e = g$ there is no solution of the system (5.30)). An interesting consequence of the inequality (5.33) and the relation (3.7) is a *lower* bound for $W^\pm$ mass (cf. on the other hand (5.12)):

$$m_W > \left(\frac{\pi\alpha}{G_F\sqrt{2}}\right)^{\frac{1}{2}} \doteq 37\,\text{GeV} \tag{5.34}$$

The inequality (5.33) is a necessary condition for the existence of a real solution of the system of equations (5.30) and it is therefore natural to call it a "condition of unification" (of weak and electromagnetic interactions) in analogy with the relation (5.11). The inequality (5.33) thus represents a condition specific for the model involving a neutral IVB. If (5.33) holds, it is easy to find out that the system (5.30) has just two solutions which differ trivially by an overall sign; however, such a difference does not lead to any physical consequences and thus we conventionally choose the solution for which (see (5.32)) $g_{WWZ} = +\sqrt{g^2 - e^2}$. Then one has

$$g_{WWZ} = \sqrt{g^2 - e^2}$$

$$g_{\nu\nu Z} = \frac{g^2}{2\sqrt{g^2 - e^2}}$$

$$g_L = \frac{-\frac{1}{2}g^2 + e^2}{\sqrt{g^2 - e^2}}$$

$$g_R = \frac{e^2}{\sqrt{g^2 - e^2}} \tag{5.35}$$

In the expressions (5.35) (similarly to (5.10)) a "unification of weak and electromagnetic interactions" is manifest, in the sense indicated at the end of Chapter 4: The coupling constants for interactions of the neutral vector boson Z are non-trivial functions of the e and g, i.e. of the parameters corresponding to the original electromagnetic and weak interaction in (4.26). Thus it seems natural to introduce the term "electroweak interactions" (which by now is standard) for such a unification of weak and electromagnetic interactions; this term was originally coined by A. Salam in 1980 and we will use it hereafter.

The solution (5.35) may also be parametrized in a somewhat different way; in view of the validity of (5.33) it is possible to introduce an angle ϑ_W (the Weinberg angle or the "weak mixing angle") such that

$$\sin \vartheta_W = \frac{e}{g} \tag{5.36}$$

and $0 < \vartheta_W < \pi/2$. The coupling constants in (5.35) may be then expressed in terms of g and ϑ_W:

$$
\begin{aligned}
g_{WWZ} &= g \cos \vartheta_W \\
g_{\nu\nu Z} &= \frac{1}{2} \frac{g}{\cos \vartheta_W} \\
g_L &= \frac{g}{\cos \vartheta_W} \left(-\frac{1}{2} + \sin^2 \vartheta_W \right) \\
g_R &= \frac{g}{\cos \vartheta_W} \sin^2 \vartheta_W
\end{aligned}
\tag{5.37}
$$

One should emphasize that the results (5.35) or (5.37) are identical with the expressions obtained for the corresponding coupling constants within the framework of the standard formulation of the GWS model (where these follow from the principle of non-abelian $SU(2) \times U(1)$ gauge invariance).

It is in order here to introduce the usual terminology: The expressions (5.15) and (5.20) obviously represent interactions of the neutral IVB with "weak neutral currents" (in contrast to the original weak interaction of charged IVB with charged currents (3.1)). As we have remarked earlier, the electromagnetic current is in this sense also neutral. In what follows we will commonly use the standard term "neutral currents" only in connection with interactions of the type (5.15) and (5.20).

Let us remark that some experimental evidence for the neutral currents was first observed in 1973; their properties predicted by the GWS theory were confirmed decisively in 1978 and repeatedly in the following years (see [42]). For some aspects of the neutral-current phenomenology see also the problem 5.16.

We shall now examine the condition (5.31) in more detail. Substituting for the coupling constants in (5.31) the corresponding expressions (5.35) or (5.37), one finds that there is indeed a *positive* solution for m_W^2/m_Z^2 (the existence of which was not obvious a priori):

$$\frac{m_W^2}{m_Z^2} = 1 - \frac{e^2}{g^2} \tag{5.38}$$

or, using (5.36)

$$\frac{m_W}{m_Z} = \cos \vartheta_W \tag{5.39}$$

The result (5.38) or (5.39) represents exactly the famous relation for the IVB masses, first derived by Weinberg [7]. The standard derivation [7] is based on an application of the Higgs mechanism [40] within the framework of the corresponding non-abelian gauge theory. In the (by now conventional) formulation [7] one has to introduce specific interactions of spin-0 fields and the relation (5.39) follows from a "minimal" realization of the Higgs mechanism (which leads to the existence of a single physical neutral scalar particle). The derivation of the relation (5.39) presented here is remarkable in that it has not been necessary to introduce any scalar particle and the corresponding interactions. From our point of view, the relation (5.39) is a consequence of the requirement of complete elimination of power-like divergences in the tree-level amplitude of the process $\nu e \to WZ$ in the limit $E \to \infty$; in particular, it follows from a condition of the compensation of some next-to-leading (linear) divergences, provided that one wants to avoid introducing physical charged spin-0 particles (see also e.g. [14]).

From (3.7), (5.36) and (5.39) one gets easily the standard formulae [7] for masses of the W and Z:

$$m_W = \left(\frac{\pi\alpha}{G_F\sqrt{2}}\right)^{\frac{1}{2}} \frac{1}{\sin\vartheta_W} \tag{5.40}$$

$$m_Z = \left(\frac{\pi\alpha}{G_F\sqrt{2}}\right)^{\frac{1}{2}} \frac{1}{\sin\vartheta_W \cos\vartheta_W} \tag{5.41}$$

The relations (5.40) and (5.41) clearly show that admissible values of IVB masses are bounded from below; we have already mentioned the lower bound for m_W earlier (see (5.34)) and from Eq. (5.41) we get one for the m_Z:

$$m_Z > 2\left(\frac{\pi\alpha}{G_F\sqrt{2}}\right)^{\frac{1}{2}} \doteq 74 \text{ GeV} \tag{5.42}$$

It should be stressed that the formulae (5.40), (5.41) give a prediction for the W and Z masses, since the parameter ϑ_W may be determined experimentally e.g. from a study of fermion scattering processes mediated by neutral current interactions of the type (5.15) and (5.20). In this context, the essential point is that one only has to know the data for relatively low energies (i.e. for $E \ll m_{IVB}$). The experimental value of the parameter $sin^2\vartheta_W$ is approximately

$$\sin^2 \vartheta_W \doteq 0.23 \tag{5.43}$$

From (5.40), (5.41) and (5.43) then follow predictions

$$m_W \doteq 77 \text{ GeV} \qquad m_Z \doteq 88 \text{ GeV} \tag{5.44}$$

The experimental determination of the $sin^2\vartheta_W$ and precise predictions for IVB masses are discussed in detail e.g. in [42] (see especially the review by R. Peccei). The experimental discovery of the particles $W^\pm$ and Z with predicted properties (see [43]) was a triumph of the GWS theory.

Let us now briefly summarize the results we have achieved so far. The starting point of our road towards a theory of electroweak interactions may be written (cf. (4.26), (4.27)) as

$$\mathcal{L}_{int} = \mathcal{L}_{CC} + \mathcal{L}_{lepton}^{(em)} + \mathcal{L}_{WW\gamma} + \mathcal{L}_{WW\gamma\gamma} + \cdots \tag{5.45}$$

where $\mathcal{L}_{CC}$ is the Lagrangian of the original weak interaction (the symbol CC stands for charged currents), the other three terms in (5.45) correspond to electromagnetic interactions and the symbol "..." represents the envisaged "missing links" of the electroweak theory. Instead of (5.45) we can now write

$$\mathcal{L}_{int} = \mathcal{L}_{CC} + \mathcal{L}_{NC} + \mathcal{L}_{lepton}^{(em)} + \mathcal{L}_{WW\gamma} + \mathcal{L}_{WWZ} + \mathcal{L}_{WW\gamma\gamma} + \cdots \tag{5.46}$$

where $\mathcal{L}_{NC}$ is the interaction of weak neutral lepton currents (i.e. the sum of expressions (5.15) and (5.20)), and the interaction term $\mathcal{L}_{WWZ}$ is given by the expression (5.13); the relevant coupling constants are given by (5.36) and (5.37). The symbol "..." in (5.46) indicates that it will be necessary to introduce further interaction terms for suppression of a "bad" high-energy behaviour of some tree-level amplitudes; for example, in the amplitude of the process $e^-e^+ \to W_L^- W_L^+$ there still remain some next-to-leading divergences, namely the terms growing linearly with $E \to \infty$. Furthermore, as we have seen in Chapter 4, severe problems with power-like growth at high energies show up in the electromagnetic contribution to the $WW \to WW$ scattering amplitude. Now we may also consider a contribution of the Z-exchange to this process. Moreover, one has to consider processes of the type $WW \to ZZ$ and $WW \to Z\gamma$ where one may expect highly divergent high-energy behaviour as well. Interactions in the sector of vector bosons are discussed in the next section.

5.4   Sector of Vector Bosons

First we shall examine in detail the tree-level scattering amplitude for $WW \to WW$. As noticed in the preceding section, in a theory with the interaction Lagrangian (5.46) one has to consider the diagrams shown in Fig. 19, as well as the electromagnetic contribution (Fig. 7). We will discuss the case where all the external W's have longitudinal polarizations. For a general WWZ interaction, contributions of the diagrams in Fig. 19 might behave like $m_W^{-4}m_Z^{-2}E^6$, since each longitudinal polarization contributes a factor of m_W^{-1} through its leading asymptotic term and the longitudinal

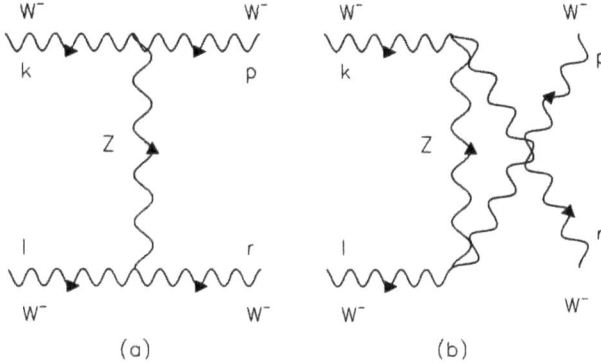

Fig. 19. Tree diagrams of the process $WW \rightarrow WW$ involving the Z exchange.

part of the Z propagator contains a factor of m_Z^{-2}. However, in Section 5.2 we have already fixed the interaction term $\mathcal{L}_{WWZ}$ in (5.46) to be of the Yang-Mills type (see (5.13), (5.14)). Using the 't Hooft identity (4.19) it is then easy to show that the contribution of the longitudinal part of the Z propagator vanishes identically even for an arbitrary combination of polarizations of the external W's. (For completeness let us add that the above-mentioned would-be leading divergence is in fact suppressed in the considered particular case $W_L W_L \rightarrow W_L W_L$ for a broader class of WWZ interactions - cf. the discussion around the relation (4.23) in Chapter 4 and see also Appendix I.) A non-trivial contribution of the diagrams in Fig. 19 thus comes only from the diagonal part of the Z propagator and the result is analogous to electromagnetic interaction (i.e. to the photon exchange in Fig. 7). For the contribution of Fig. 19 one may thus write

$$\mathcal{M}_a^{(Z)} + \mathcal{M}_b^{(Z)} = g_{WWZ}^2 \frac{1}{4m_W^4}(t^2 + u^2 - 2s^2) + O(E^2) + O(1) \qquad (5.47)$$

(cf. (4.24) and Appendix J). Within the framework of a provisional theory described by the Lagrangian (5.46), the full tree-level amplitude for $W_L W_L \rightarrow W_L W_L$ is of course obtained by summing the electromagnetic and Z-exchange contributions, i.e., it is given by the sum of (4.24) and (5.47). Using the first of the relations (5.35) (see also Eq. (5.32)) one gets for the full contribution of Fig. 7 and Fig. 19

$$\mathcal{M}^{(\gamma, Z)} = g^2 \frac{1}{4m_W^4}(t^2 + u^2 - 2s^2) + O(E^2) + O(1) \qquad (5.48)$$

Now it is obvious that within a model described by (5.46) the leading quartic divergence in (5.48) could be eliminated only by a trivial choice $g = 0$ (which is unacceptable). Thus we must add new interactions to the terms already present in (5.46), which would give a non-trivial tree-level contribution to the scattering amplitude of $W_L W_L \rightarrow W_L W_L$, diverging as E^4 in the high-energy limit and cancelling the leading divergence in (5.48). It is not difficult to realize that the only possibility is to introduce a direct self-interaction of four vector fields W (an interaction of vector bosons with a scalar field is of no use here, as it is not sufficient for the suppression of quartic divergences). Imposing the constraint (5.5), it is clear that terms involving derivatives of vector fields are not admissible. The most general interaction of the required type must obviously have the form

$$\mathcal{L}_{WWWW} = a(W_\mu^- W^{+\mu})(W_\nu^- W^{+\nu}) + b(W_\mu^- W^{-\mu})(W_\nu^+ W^{+\nu}) \qquad (5.49)$$

where a and b are real constants. In the first order of perturbation expansion the interaction (5.49) yields a contribution to the scattering amplitude of the process $W_L W_L \rightarrow W_L W_L$, which for $E \rightarrow \infty$ (see the problem 5.3) may be written as

$$\mathcal{M}^{(4W)} = a\frac{1}{2m_W^4}(t^2 + u^2) + b\frac{1}{m_W^4}s^2 + O(E^2) + O(1) \qquad (5.50)$$

Now it is obvious that the leading high-energy divergences in Eq. (5.48) and Eq. (5.50) mutually cancel if and only if

$$a = -\frac{1}{2}g^2, \qquad b = \frac{1}{2}g^2 \qquad (5.51)$$

and the sought Lagrangian for a direct interaction of four W's thus has the form

$$\mathcal{L}_{WWWW} = \frac{1}{2}g^2(W^-)^2(W^+)^2 - \frac{1}{2}g^2(W^- . W^+)^2 \qquad (5.52)$$

(see (5.49), (5.51)). In (5.52) we of course use the standard shorthand notation for a Lorentz scalar product and for the square of a four-vector; such a notation will be used frequently in similar expressions in what follows. It is interesting to notice that coupling constants in the contact interaction of four W's (5.52) are proportional to g^2; one should keep in mind that the g is originally the coupling constant for the interaction of the W with charged fermion currents (which do not play any role in the considered process $WW \rightarrow WW$). This remarkable and at first sight rather unexpected correspondence between two completely different interactions is of course a technical consequence of repeated application of divergence cancellation conditions for tree-level scattering amplitudes of several distinct processes. Within the framework of the traditional approach such relations arise naturally from the structure of non-abelian gauge theory (see e.g. [25] etc.).

Tree-level Feynman diagrams of the process $WW \rightarrow WW$ in Fig. 7, Fig. 19 and the diagram corresponding to the contact interaction (5.52) are collected in Fig. 20. For the full contribution of these diagrams (involving longitudinally polarized W's) one gets after a rather tedious calculation (see the problem 5.4 and Appendix J) the result

$$\mathcal{M}_a + \mathcal{M}_b + \mathcal{M}_c = -g^2 \frac{s}{4m_W^2} + O(1) \tag{5.53}$$

It is obvious that the remaining divergence in (5.33) cannot be eliminated without adding a new term to the interaction Lagrangian; taking into account that this divergence is only quadratic, one could attempt to compensate it by means of an additional diagram involving the exchange of a (neutral) spin-0 particle, i.e. by introducing a new interaction of the vector field W with a scalar field. We have encountered an analogous problem in the preceding section for a different process (cf. the discussion following the relation (5.29)). The problem of suppressing such "residual" divergences in Eq. (5.53) and in the other tree-level amplitudes will be treated in detail in Section 5.5.

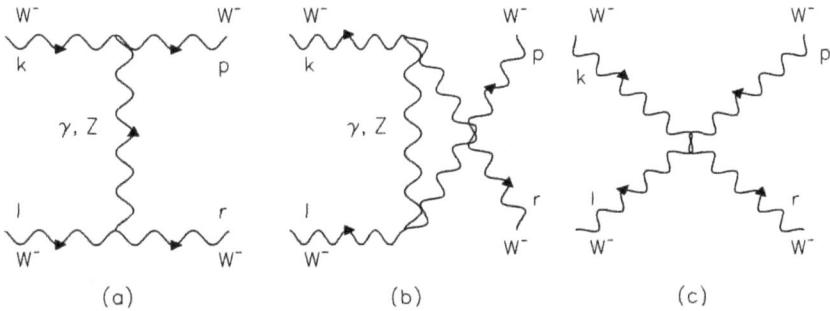

Fig. 20. Tree-level diagrams for $WW \to WW$ corresponding to the trilinear interactions $WW\gamma$, WWZ and the direct contact interaction $WWWW$.

We will now discuss other binary processes in the sector of vector bosons, i.e. processes of the type $V_1 V_2 \to V_3 V_4$, where V_i, $i = 1, ..., 4$ generally denote $W^\pm$, Z or γ. If we take into account the interactions introduced up to now, then on the tree level there occur only processes $WW \to \gamma\gamma$, $WW \to ZZ$ and $WW \to Z\gamma$ (the first of them has been discussed in detail in Chapter 4). First let us consider the process $W^- W^+ \to ZZ$. Relevant tree diagrams are shown in Fig. 21. We will consider again the case where all four external vector bosons have longitudinal polarizations. As for the diagrams in Fig. 19 it is easy to show that the leading (quartic) divergence comes only from the contribution of the diagonal term in the W propagator. In the high-energy limit we then obtain for diagrams in Fig. 21

$$\mathcal{M}_a + \mathcal{M}_b = -\frac{1}{4}g_{WWZ}^2 \frac{1}{m_W^2 m_Z^2}(t^2 + u^2 - 2s^2) + O(E^2) + O(1) \tag{5.54}$$

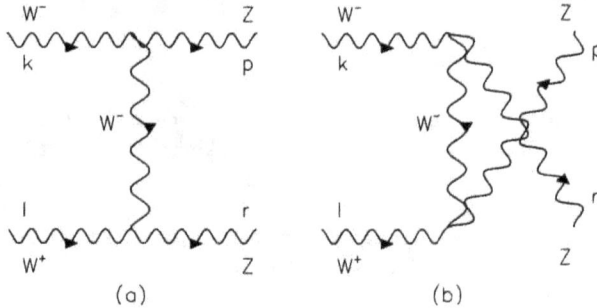

Fig. 21. Tree-level diagrams for the process $W^- W^+ \to ZZ$ arising from the trilinear interaction WWZ.

where the coupling constant g_{WWZ} is of course given by Eq. (5.35) (or Eq. (5.37)). For a compensation of the leading divergence in Eq. (5.54) we introduce a new contact interaction of four vector fields

$$\mathcal{L}_{WWZZ} = c(W_\mu^- Z^\mu)(W_\nu^+ Z^\nu) + d(W_\mu^- W^{+\mu})(Z_\nu Z^\nu) \qquad (5.55)$$

where c, d are real constants; the option (5.55) obviously represents the most general interaction Lagrangian with required properties. In the first order of perturbation expansion the interaction (5.55) gives rise to the Feynman diagram shown in Fig. 22. For the contribution of this graph in the limit $E \to \infty$ (see the problem 5.5) one then gets

$$\mathcal{M}^{WWZZ} = \frac{1}{4} \frac{1}{m_W^2 m_Z^2} [c(t^2 + u^2) + 2ds^2] + O(E^2) + O(1) \qquad (5.56)$$

The condition of mutual compensation of leading divergences in the expressions (5.54) and (5.56) is thus equivalent to

$$c = g_{WWZ}^2, \qquad d = -g_{WWZ}^2 \qquad (5.57)$$

Thus we have fixed another piece of the necessary direct interaction of four vector bosons, namely

$$\mathcal{L}_{WWZZ} = g_{WWZ}^2 [(W^- . Z)(W^+ . Z) - (W^- . W^+)Z^2] \qquad (5.58)$$

However, introducing the interaction term (5.58) is not enough to suppress quadratic divergences in the tree-level amplitude of $W_L^- W_L^+ \to Z_L Z_L$; as in all previous cases, we defer this problem to Section 5.5.

Finally we shall examine the scattering amplitude of the process $W^- W^+ \to Z\gamma$. The relevant tree diagrams corresponding to trilinear interactions of the vector fields

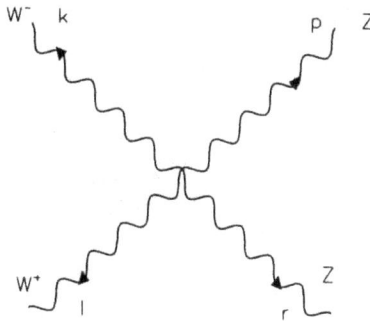

Fig. 22. *The lowest-order diagram for* $W^-W^+ \to ZZ$ *corresponding to the direct interaction of four vector fields.*

are shown in Fig. 23. As in the preceding cases let us consider a configuration in which all massive vector bosons $W^\pm$ and Z have longitudinal polarizations. The leading divergent term appearing in the corresponding scattering amplitude for $E \to \infty$ then behaves like $m_W^{-2} m_Z^{-1} E^3$ and comes from the diagonal part of the W propagator; its longitudinal part may only contribute to a next-to-leading (linear) divergence, as one may easily find by means of the 't Hooft identity (4.19). A direct evaluation of the diagrams in Fig. 23 leads to the result

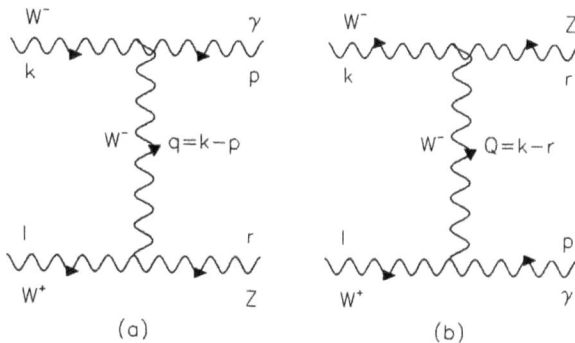

Fig. 23. *Tree-level diagrams of the process* $W^-W^+ \to Z\gamma$ *arising from trilinear interactions WWZ and WWγ.*

$$\mathcal{M}_a + \mathcal{M}_b = g_{WW\gamma}g_{WWZ}\frac{1}{m_W^2 m_Z}[s(r.\varepsilon^*(p)) - (l.r)(k.\varepsilon^*(p)) - (k.r)(l.\varepsilon^*(p))]$$
$$+ O(E) + O(1) \tag{5.59}$$

where we have used the symbol $g_{WW\gamma}$ for the electromagnetic coupling constant e and $\varepsilon(p)$ stands for a photon polarization (which is transverse, of course). To compensate the leading divergence in Eq. (5.59) we have to introduce another contact interaction of the four vector bosons $W^\pm$, Z and γ; the most general form of such an interaction satisfying the usual requirements is

$$\begin{aligned} \mathcal{L}_{WWZ\gamma} = & f(W_\mu^- W^{+\mu})(Z_\nu A^\nu) \\ & + g(W_\mu^- Z^\mu)(W_\nu^+ A^\nu) \\ & + h(W_\mu^- A^\mu)(W_\nu^+ Z^\nu) \end{aligned} \tag{5.60}$$

where f, g and h are real constants.

In the first order of perturbation expansion the interaction (5.60) yields the Feynman diagram shown in Fig. 24. Its contribution to the scattering amplitude of the

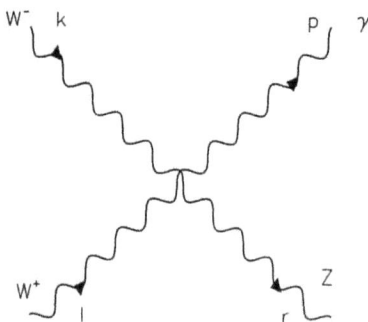

Fig. 24. *The lowest-order diagram of the process* $W^- W^+ \to Z\gamma$ *corresponding to a direct interaction of the four vector fields.*

process $W_L^- W_L^+ \to Z_L\gamma$ may be written in the high-energy limit as

$$\mathcal{M}^{(WWZ\gamma)} = \frac{1}{m_W^2 m_Z}\left[\frac{1}{2}fs(r.\varepsilon^*(p)) + g(k.r)(l.\varepsilon^*(p)) + h(l.r)(k.\varepsilon^*(p))\right]$$
$$+ O(E) + O(1) \tag{5.61}$$

Leading divergences in Eq. (5.59) and Eq. (5.61) thus cancel each other if and only if

$$f = -2g_{WW\gamma}g_{WWZ}, \qquad g = h = g_{WW\gamma}g_{WWZ} \tag{5.62}$$

The needed "compensating" direct interaction $WWZ\gamma$ is thus described by the Lagrangian

$$\mathcal{L}_{WWZ\gamma} = g_{WW\gamma}g_{WWZ}[-2(W^-.W^+)(A.Z) + (W^-.Z)(W^+.A)$$
$$+ (W^-.A)(W^+.Z)] \tag{5.63}$$

It can be shown that by adding Fig. 24 to the diagrams in Fig. 23 the non-leading high-energy divergences are in fact cancelled as well; more detailed comments on this remarkable fact will be given in the next section.

The following remark concerning the interaction $WW\gamma\gamma$ is also in order here: In Chapter 4 we obtained a direct interaction of this type automatically as a part of the $U(1)$ gauge invariant electromagnetic interaction of charged vector bosons $W^\pm$ (see (4.8), (4.10)); from the considerations presented in this section it is clear that the corresponding term $\mathcal{L}_{WW\gamma\gamma}$ could also be derived from the requirement of divergence cancellations in the diagrams shown in Fig. 6.

Our results concerning the direct (contact) interactions of four vector bosons $W^\pm, Z$ or γ may be summarized as follows: According to (4.10), (5.52), (5.58), (5.63) and using the relations (5.36), (5.37) we have

$$\mathcal{L}_{WW\gamma\gamma} = -g^2 \sin^2 \vartheta_W \left[(W^-.W^+)A^2 - (W^-.A)(W^+.A)\right]$$
$$\mathcal{L}_{WWWW} = \frac{1}{2}g^2 \left[(W^-)^2(W^+)^2 - (W^-.W^+)^2\right]$$
$$\mathcal{L}_{WWZZ} = -g^2 \cos^2 \vartheta_W \left[(W^-.W^+)Z^2 - (W^-.Z)(W^+.Z)\right]$$
$$\mathcal{L}_{WWZ\gamma} = g^2 \sin\vartheta_W \cos\vartheta_W[-2(W^-.W^+)(A.Z) + (W^-.Z)(W^+.A) +$$
$$+ (W^-.A)(W^+.Z)] \tag{5.64}$$

The expressions (5.64) may conveniently be rewritten in the following compact form: Denoting by $\mathcal{L}_{VVVV}$ the sum

$$\mathcal{L}_{VVVV} = \mathcal{L}_{WWWW} + \mathcal{L}_{WW\gamma\gamma} + \mathcal{L}_{WWZZ} + \mathcal{L}_{WWZ\gamma}, \tag{5.65}$$

then

$$\mathcal{L}_{VVVV} = -g^2[\frac{1}{2}(W^-.W^+)^2 - \frac{1}{2}(W^-)^2(W^+)^2 + (W^0)^2(W^-.W^+) -$$
$$- (W^-.W^0)(W^+.W^0)], \tag{5.66}$$

where we have also introduced a new shorthand notation for the relevant combination of neutral vector fields:

$$W^0_\mu = \cos\vartheta_W Z_\mu + \sin\vartheta_W A_\mu \tag{5.67}$$

Now it is also possible to recast the trilinear interactions of vector bosons in a more compact form; defining $\mathcal{L}_{VVV}$ as the sum

$$\mathcal{L}_{VVV} = \mathcal{L}_{WW\gamma} + \mathcal{L}_{WWZ} \tag{5.68}$$

then using (4.11), (5.13), (5.36), (5.37) and the definition (5.67) one has

$$\mathcal{L}_{VVV} = -ig(W^{0\mu}W^{-\nu}\overleftrightarrow{\partial}_{\mu}W_{\nu}^{+} + W^{-\mu}W^{+\nu}\overleftrightarrow{\partial}_{\mu}W_{\nu}^{0} + W^{+\mu}W^{0\nu}\overleftrightarrow{\partial}_{\mu}W_{\nu}^{-}) \tag{5.69}$$

Thus instead of the interaction Lagrangian (5.46) one may write

$$\mathcal{L}_{int} = \mathcal{L}_{CC} + \mathcal{L}_{NC} + \mathcal{L}_{lepton}^{(em)} + \mathcal{L}_{VVV} + \mathcal{L}_{VVVV} + \dots \tag{5.70}$$

The symbol "..." in (5.70) means the remaining "missing links", i.e. the interaction terms which we will have to introduce for a compensation of non-leading high-energy divergences which still occur in some tree-level amplitudes, as e.g. in Eq. (5.53) etc. These residual divergences and their elimination are the subject of the next section.

5.5 Residual Divergences and Scalar Boson

Let us return to the formula (5.53) which expresses the contribution to scattering amplitude of the process $W_L W_L \to W_L W_L$ corresponding to the diagrams in Fig. 20. As indicated in the preceding section, we will now try to eliminate the remaining quadratic divergence in Eq. (5.53) by introducing a new interaction of the W's with a neutral scalar field (which we denote here by η). It is not difficult to realize that the only possible choice (satisfying our standard requirements) is represented by the interaction Lagrangian

$$\mathcal{L}_{WW\eta} = g_{WW\eta}W_{\mu}^{-}W^{+\mu}\eta \tag{5.71}$$

(cf. also (5.6)). Tree diagrams for the process $W^{-}W^{-} \to W^{-}W^{-}$ corresponding to the interaction (5.71) are shown in Fig. 25. As stated earlier, the coupling constant $g_{WW\eta}$ in (5.71) must have the dimension of mass (see (5.8)). Then it is also obvious that the contribution of diagrams in Fig. 25 to the scattering amplitude of $W_L^- W_L^- \to W_L^- W_L^-$ may involve at most quadratic divergence in the limit $E \to \infty$. Indeed, a corresponding asymptotic term may be estimated in this case as $g_{WW\eta}^2 m_W^{-4} E^2$. Direct evaluation of the diagrams in Fig. 25 for longitudinally polarized W's (see the problem 5.7) leads to the result

$$\mathcal{M}_a^{(\eta)} + \mathcal{M}_b^{(\eta)} = g_{WW\eta}^2 \frac{s}{m_W^4} + O(1) \tag{5.72}$$

From (5.53) and (5.72) it is clear that the desired cancellation of residual quadratic divergences in the scattering amplitude of $W_L^- W_L^- \to W_L^- W_L^-$ occurs if and only if

$$g_{WW\eta} = gm_W \tag{5.73}$$

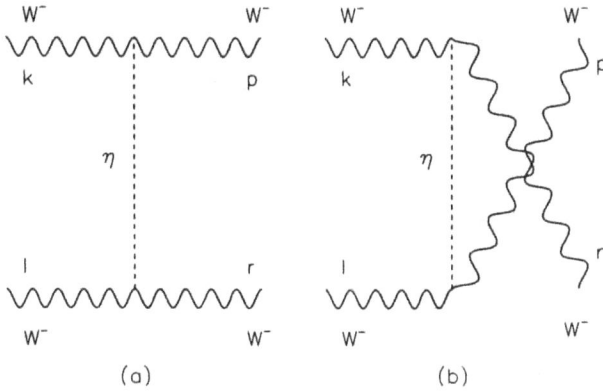

Fig. 25. Tree-level diagrams of the process $W^-W^- \to W^-W^-$ involving the exchange of a scalar boson η.

This result is another remarkable example of the fact that offending high-energy divergences arising in the individual diagrams may indeed be cancelled in the full tree-level scattering amplitude if the relevant coupling constants are judiciously chosen; at the same time it is also obvious that within our "minimal strategy" such a choice is essentially unique. Eq. (5.73) represents a new non-trivial relation among coupling constants and masses in different sectors of the model we are building; the existence of many such relations is a typical feature of the theory of electroweak unification.

From what we have said earlier in this chapter it is obvious that the new interaction term (5.71) will also play an important role in scattering amplitudes of some other binary processes. In particular, we shall now return to the process $e^-e^+ \to W_L^- W_L^+$. For the total contribution of the diagrams that we have considered up to now (see Fig. 17) we obtain (using (5.21), (5.22), (5.23), (J.1), (J.5) and (5.36), (5.37)) the expression

$$\mathcal{M}_a + \mathcal{M}_b + \mathcal{M}_c = -\frac{g^2}{4m_W^2} m\bar{v}(l)u(k) + O(1) \tag{5.74}$$

(where the relevant four-momenta are of course denoted according to Fig. 17). It is interesting to notice that terms proportional to $\bar{v}(l)\gamma_5 u(k)$, occurring in the individual diagrams (a) and (c) (see (J.1), (J.5)) cancel in their sum as a consequence of the relations (5.36), (5.37). Now we may try to eliminate the remaining linear divergence in the expression (5.74) by means of a "compensation" diagram involving an exchange of the scalar boson η which we have already discussed briefly in Section 5.2 (see Fig. 12 and the considerations following the relation (5.8)). Of course, for this purpose one also has to introduce an interaction of $e^\pm$ with the scalar field η; from the structure of the residual linear divergence in (5.74) it is seen that it is sufficient to consider the "Yukawa interaction" (cf. (5.7))

$$\mathcal{L}_{ee\eta} = g_{ee\eta}\bar{e}e\eta \tag{5.75}$$

For longitudinally polarized W's one then gets a result for the contribution of Fig. 12 (see the problem 5.8) which in the high-energy limit may be written as

$$\mathcal{M}^{(\eta)} = -\frac{1}{2m_W^2}g_{ee\eta}g_{WW\eta}\bar{v}(l)u(k) + O(1) \tag{5.76}$$

where $g_{WW\eta}$ is of course defined by (5.73). Required cancellation of the linearly divergent terms in the sum of (5.74) and (5.76) then occurs if and only if

$$g_{ee\eta} = -\frac{g}{2}\frac{m}{m_W} \tag{5.77}$$

The results (5.73) and (5.77)) reflect one remarkable common feature of trilinear interactions of the scalar field η: A coupling constant is always proportional to the mass of the particle interacting with the η. Within our approach, such a dependence is obviously related to the fact that interactions involving the scalar field are introduced to compensate *non-leading* high-energy divergences, which in comparison with leading terms contain extra factors of M or M^2 where M is a mass. (Let us remark that within the framework of a gauge theory of electroweak interactions a simple alternative interpretation of the above relations follows naturally from the Higgs mechanism, which generates masses of vector bosons and fermions; this traditional formulation can be found in any standard textbook or monograph - see e.g. [17], [21], [25] etc.)

We shall now examine other binary processes for which there are still power-like high-energy divergences in the corresponding scattering amplitudes. In Section 5.3 we mentioned that the tree-level scattering amplitude for $e^+e^- \to Z_L Z_L$ contains a linear divergence if one takes into account only the diagrams shown in Fig. 26(a), (b). Indeed, a direct computation of the diagrams (a), (b) (see the problem 5.9) gives the result

$$\mathcal{M}_a + \mathcal{M}_b = -(g_L - g_R)^2\frac{m}{m_Z^2}\bar{v}(l)u(k) + O(1) \tag{5.78}$$

where g_L, g_R are coupling constants for the interaction of the Z and neutral currents, given by the corresponding expressions (5.35) or (5.37). Let us emphasize that quadratic divergences contained in the individual diagrams (a) and (b) automatically cancel in their sum (the very existence of the crossed graph (b) is of course due to the fact that Z is neutral); such an effect is in a sense analogous to the mechanism of divergence cancellations in the electrodynamics with a "heavy photon"   cf. the problem 3.7. We may now try to compensate the linear divergence in (5.78) by means of the diagram (c) in Fig. 26. One vertex of this diagram corresponds to the interaction (5.75) while an appropriate interaction producing the other vertex has yet to be introduced. It is clear that in analogy with (5.71) one may write for the corresponding Lagrangian generally

$$\mathcal{L}_{ZZ\eta} = g_{ZZ\eta} Z_\mu Z^\mu \eta \tag{5.79}$$

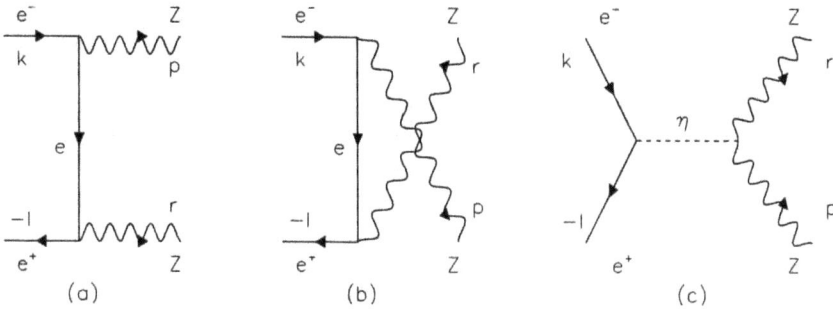

Fig. 26: Tree-level diagrams for $e^+e^- \to ZZ$.

For the contribution of the diagram (c) in Fig. 26 one then easily gets (see the problem 5.10)

$$\mathcal{M}_c = -g_{een} g_{ZZ\eta} \frac{1}{m_Z^2} \bar{v}(l) u(k) + O(1) \tag{5.80}$$

where the coupling constant is of course given by (5.77). Using (5.37) one gets

$$(g_L - g_R)^2 = \frac{g^2}{4 \cos^2 \vartheta_W}, \tag{5.81}$$

and thus finds immediately that linear divergences in (5.78) and (5.80) cancel each other if and only if

$$g_{ZZ\eta} = \frac{1}{2 \cos \vartheta_W} g m_Z \tag{5.82}$$

(in deriving (5.82) we have also used the relation $m_W = m_Z \cos \vartheta_W$ - see (5.39)).

Let us note that the interaction term (5.79) we have just introduced should now also lead to a compensation of residual quadratic divergences e.g. in the scattering amplitude of the process $W_L^- W_L^+ \to Z_L Z_L$ discussed in the preceding section (cf. the considerations following eq. (5.58)); more precisely, such an automatic cancellation of divergences would be highly desirable in order not to have to introduce further interaction terms. One may verify directly that the above-mentioned elimination of quadratic divergences indeed occurs. However, the corresponding (rather tedious) calculation will not be performed here; instead we shall comment on this remarkable fact from a more general point of view later in this section.

Introduction of the scalar field η and the corresponding interactions may of course lead to new power-like divergences in the limit $E \to \infty$, i. e. one may anticipate divergent terms in tree-level scattering amplitudes of processes which we have not considered so far. Indeed, one also has to investigate processes involving real scalar bosons in the initial or final state (note that in the diagrams considered up to now the η always entered as a virtual exchanged particle); it is clear that for the tree diagrams involving external lines of scalar bosons and massive vector bosons one may in general expect - as a consequence of the by now familiar mechanisms - a divergent behaviour in the limit $E \to \infty$. In particular, we shall now examine the process of production of a pair of scalar bosons in the annihilation of a pair of longitudinally polarized W's, i. e. the process $W_L^- W_L^+ \to \eta\eta$. In such a case the interaction term (5.71) leads to the tree diagrams shown in Fig. 27 (a), (b). For the total contribution of the diagrams (a), (b) in the limit $E \to \infty$ one may then write (see the problem 5.11)

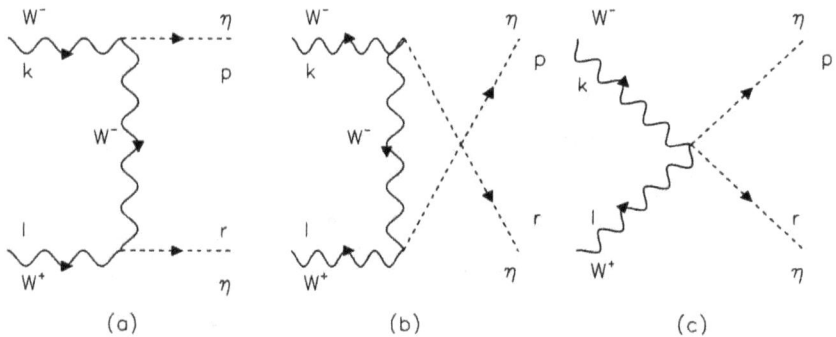

Fig. 27. *Tree-level diagrams of the process* $W^- W^+ \to \eta\eta$.

$$\mathcal{M}_a + \mathcal{M}_b = -\frac{g^2}{4m_W^2}s + O(1) \tag{5.83}$$

For a compensation of the quadratic divergence in (5.83) one has to introduce a new interaction term; obviously, the only possibility (satisfying the usual conditions) is represented by the expression

$$\mathcal{L}_{WW\eta\eta} = g_{WW\eta\eta} W_\mu^- W^{+\mu} \eta^2 \tag{5.84}$$

which in the lowest order of perturbation expansion produces the diagram in Fig. 27(c). In the case of longitudinally polarized $W^\pm$ it is easy to get for the contribution of this diagram

$$\mathcal{M}_c = g_{WW\eta\eta} \frac{1}{m_W^2}s + O(1) \tag{5.85}$$

The requirement of divergence cancellation between Eq. (5.83) and Eq. (5.85) is therefore equivalent to

$$g_{WW\eta\eta} = \frac{1}{4}g^2 \qquad (5.86)$$

Similarly one may consider the process $Z_L Z_L \to \eta\eta$; for a compensation of quadratic divergence in tree-level diagrams descending from the trilinear interaction (5.79) it is necessary to introduce a direct contact interaction $ZZ\eta\eta$,

$$\mathcal{L}_{ZZ\eta\eta} = g_{ZZ\eta\eta} Z_\mu Z^\mu \eta^2 \qquad (5.87)$$

and the requirement of divergence cancellation in the corresponding diagrams (which can be obtained from Fig. 27 by replacing all W's with Z's) yields, using (5.82), the following relation for the coupling constant $g_{ZZ\eta\eta}$:

$$g_{ZZ\eta\eta} = \frac{1}{8} \frac{g^2}{\cos^2 \vartheta_W} \qquad (5.88)$$

Now it is in order to summarize briefly the results we have obtained so far. By means of a systematic elimination of high-energy power-like divergences in tree-level scattering amplitudes of some selected binary processes we have arrived at the interaction Lagrangian

$$\begin{aligned} \mathcal{L}_{int} &= \mathcal{L}_{CC} + \mathcal{L}_{NC} + \mathcal{L}_{lepton}^{(em)} + \mathcal{L}_{VVV} + \mathcal{L}_{VVVV} + \mathcal{L}_{WW\eta} \\ &+ \mathcal{L}_{ZZ\eta} + \mathcal{L}_{WW\eta\eta} + \mathcal{L}_{ZZ\eta\eta} + \mathcal{L}_{ee\eta} + ... \end{aligned} \qquad (5.89)$$

(see (5.70), (5.71), (5.75), (5.79), (5.84) and (5.87)) where the coupling constants of the newly introduced interactions (i.e. of those which are new with respect to (4.26)) are intertwined via many remarkable relations (see (5.36), (5.37), (5.73), (5.77) etc.). The symbol "..." again denotes further possible terms which should eventually be included for the theory of electroweak interactions to be complete (i.e. so that it would satisfy the condition (5.4) or at least the tree-unitarity criterion (5.1) in all cases). It should be emphasized that some terms, which *a priori* are not excluded by the requirement of Lorentz invariance and by the condition $\dim \mathcal{L}_{int} \leq 4$ (see (5.5)), are manifestly absent in the Lagrangian (5.89). For example, in the $\mathcal{L}_{VVV}$ there is no ZZZ term and similarly the $\mathcal{L}_{VVVV}$ does not incorporate any term of the type $ZZZZ$. Within our approach, the absence of such terms is related to the fact that for some processes certain divergences cancel automatically (e.g. for $e^- e^+ \to Z_L Z_L$ or $e^- e^+ \to Z_L \gamma$ - see (5.78) and the problem 5.12) and the above-mentioned "exotic" interaction terms are simply not necessary. (The absence of a $ZZ\gamma$ term is of course also completely natural from the physical point of view, as its existence would mean a direct electromagnetic interaction of the neutral Z.) Potentially interesting (i.e. potentially "dangerous") binary processes which we have not considered in detail and the corresponding tree-level scattering amplitudes will be discussed later in this section; the above remarks are pointing toward a preliminary conclusion that for the elimination of unacceptable high-energy behaviour of tree diagrams of *binary* processes the terms given explicitly in the Lagrangian (5.89) are sufficent.

However, we are still not at the end of our road. We may also consider the remaining two interaction terms of renormalizable type (i.e. satisfying the condition (5.5)), namely a cubic and a quartic self-interaction of the neutral scalar field η:

$$\mathcal{L}_{\eta\eta\eta} = g_{\eta\eta\eta}\eta^3 \tag{5.90}$$

$$\mathcal{L}_{\eta\eta\eta\eta} = g_{\eta\eta\eta\eta}\eta^4 \tag{5.91}$$

(Let us recall that the coupling constant in (5.90) then has dimension of a mass, while the coupling constant in (5.91) is dimensionless.) Obviously, the interaction terms (5.90) and (5.91) need not be introduced for a compensation of power-like high-energy divergences in tree-level scattering amplitudes of binary processes. However, they play an important role in some processes of the type $1 + 2 \to 3 + 4 + 5$ (this remarkable fact was first noticed by Cornwall, Levin and Tiktopoulos [11]). The corresponding calculations are technically rather complicated, so here we only recapitulate the essential results [11] very briefly. First one has to recall generally that for a process involving 5 particles the corresponding scattering amplitude has dimension $[M^{-1}]$ in units of an arbitrary mass (see (C.3)) and the condition of tree unitarity requires for the five-point amplitude a high-energy behaviour

$$\mathcal{M}_{1+2\to3+4+5} \simeq \frac{1}{E} \tag{5.92}$$

(see (5.3)), where E is a typical energy of the considered process (e.g. $E = \sqrt{s}$). In the paper [11] (cf. also [39]) the processes $ZZ \to ZZ\eta$ and $ZZ \to \eta\eta$ (and also the corresponding processes involving charged vector bosons) were investigated from such a point of view. Basic types of tree-level diagrams contributing to the scattering amplitudes of these processes are shown in Fig. 28 and 29. (As an instructive exercise we recommend the reader to draw all the tree diagrams derived from the basic types in Fig. 28, 29 and verify that e.g. in the case of the process $ZZ \to \eta\eta$ the total number of graphs is 25.)

In what follows we discuss only the case of longitudinally polarized Z's in the considered processes. Then the diagrams of the type (a), (b) in Fig. 28 (i.e. those in which the cubic self-interaction (5.90) is not involved) give a contribution whose leading term behaves in the limit $E \to \infty$ as a constant independent of E; this asymptotically constant term (coming from the diagrams of the type (a)) may be estimated (up to a numerical factor) as $m_Z^{-4}m_Z^{-2}g_{ZZ\eta}^3m_\eta^2$, where $g_{ZZ\eta}$ is the coupling constant (5.82). The contribution of diagrams of the type (c) (i.e. those which involve the self-interaction (5.90)) also contains an asymptotically constant term which may be estimated (up to a factor) as $m_Z^{-4}g_{ZZ\eta}^2g_{\eta\eta\eta}$. Since the whole tree-level amplitude of the process $Z_LZ_L \to Z_LZ_L\eta$ should exhibit the "good" high-energy behaviour (5.92), one has to achieve a cancellation of the above-mentioned asymptotically constant terms by means of an appropriate choice of the coupling constant $g_{\eta\eta\eta}$. An explicit

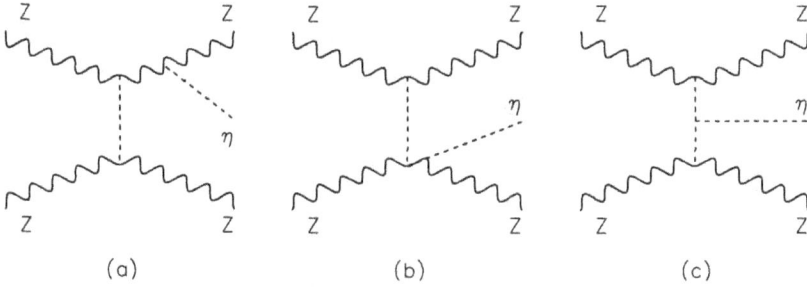

Fig. 28. *Basic types of tree-level diagrams of the process* $ZZ \to ZZ\eta$. *All the other graphs correspond to appropriate (topologically distinct) permutations of external lines and vertices.*

calculation [11] then leads to the conclusion that desired cancellation of unwanted constant terms occurs if and only if

$$g_{\eta\eta\eta} = -\frac{1}{4} g \frac{m_\eta^2}{m_W} \tag{5.93}$$

It is interesting to notice, among other things, that in this connection it was necessary to consider explicitly for the first time a non-zero mass of the neutral scalar boson η, i.e. the parameter $m_\eta \neq 0$. For the process $Z_L Z_L \to \eta\eta\eta$ the condition of a compensation of asymptotically constant contributions from diagrams of the type (a) – (e) in Fig. 29 by similar terms coming from graphs of the type (f) (i.e. from those involving the quartic self-interaction (5.91)) amounts to fixing the coupling constant $g_{\eta\eta\eta\eta}$:

$$g_{\eta\eta\eta\eta} = -\frac{1}{32} g^2 \frac{m_\eta^2}{m_W^2} \tag{5.94}$$

Similarly to (5.93), it is essential here that $m_\eta \neq 0$; however, the preceding considerations do not imply any constraint for the value of m_η (in contrast to masses of vector bosons $W^\pm$ and Z which have been accurately predicted by the theory of electroweak unification - see (5.40), (5.41)).

Let us summarize the results which we have obtained up to now in constructing a theory of electroweak interactions. We have arrived at an interaction Lagrangian which has the form (cf. (5.89))

$$\begin{aligned} \mathcal{L}_{int} &= \mathcal{L}_{CC} + \mathcal{L}_{NC} + \mathcal{L}_{lepton}^{(em)} + \mathcal{L}_{VVV} + \mathcal{L}_{VVVV} + \mathcal{L}_{WW\eta} \\ &+ \mathcal{L}_{ZZ\eta} + \mathcal{L}_{WW\eta\eta} + \mathcal{L}_{ZZ\eta\eta} + \mathcal{L}_{ee\eta} + \mathcal{L}_{\eta\eta\eta} + \mathcal{L}_{\eta\eta\eta\eta} + ... \end{aligned} \tag{5.95}$$

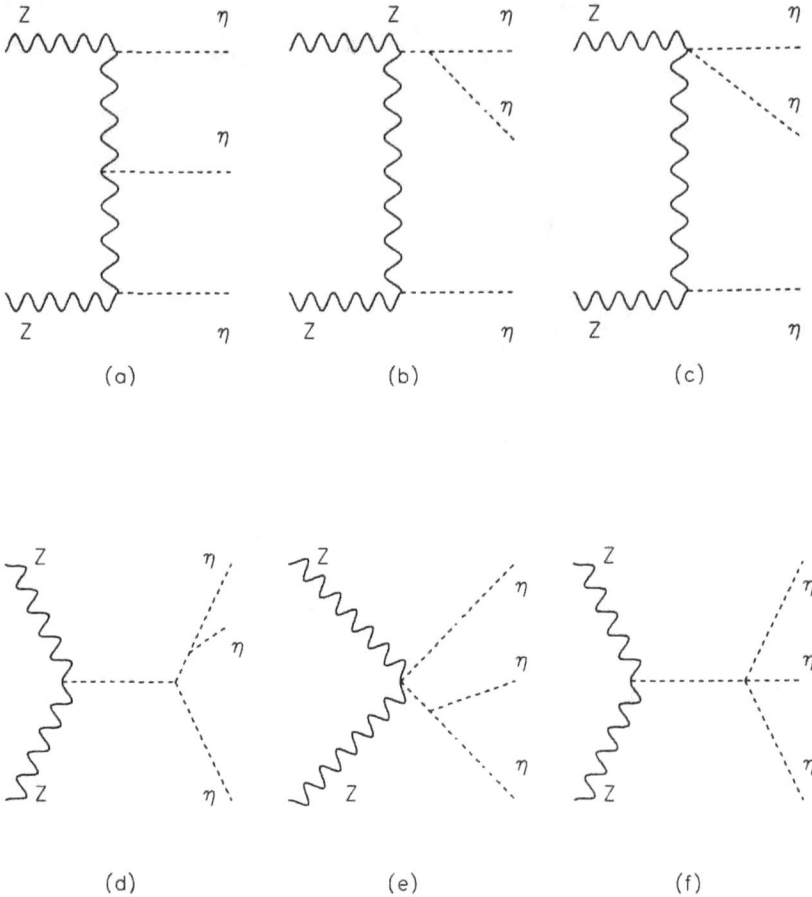

Fig. 29. *Basic types of tree-level diagrams of the process* $ZZ \to \eta\eta\eta$. *All the other graphs are obtained by appropriate (topologically distinct) permutations of external lines and vertices.*

where the last two terms in (5.95) are given by the expressions (5.90), (5.91), (5.93) and (5.94). The interaction terms written explicitly in (5.95) are *necessary* for suppressing the bad high-energy behaviour of individual tree-level diagrams corresponding to the given model. For fixing the corresponding coupling constants we have employed only a limited number of physical scattering processes and in the construction (5.95) we have used up all the interaction terms which had to be taken into account within the framework of the "minimal" strategy adopted. More precisely, we have employed tree-level scattering amplitudes of the following processes (in each case for a particular combination of helicities of the incoming and outgoing particles):

$$
\begin{aligned}
\nu\bar{\nu} &\rightarrow W^- W^+ \\
e^- e^+ &\rightarrow W^- W^+ \\
\bar{\nu} e &\rightarrow W^- Z \\
e^- e^+ &\rightarrow Z Z \\
W^- W^+ &\rightarrow \gamma\gamma \\
W^- W^- &\rightarrow W^- W^- \\
W^- W^+ &\rightarrow Z Z \\
W^- W^+ &\rightarrow Z\gamma \\
W^- W^+ &\rightarrow \eta\eta \\
Z Z &\rightarrow \eta\eta \\
Z Z &\rightarrow Z Z \eta \\
Z Z &\rightarrow \eta\eta\eta
\end{aligned}
\tag{5.96}
$$

It is not clear *a priori* whether the desirable divergence cancellations also occur in the tree-level scattering amplitudes of the other physical processes which we have not considered yet (this uncertainty is expressed by the symbol "..." in (5.95)). In particular, if we restrict ourselves to binary processes, then besides (5.96) there are several other cases which are potentially interesting (i.e. potentially "dangerous") from the point of view of the high-energy behaviour of the relevant tree diagrams, namely

$$
\begin{aligned}
\bar{\nu} e &\rightarrow W^- \eta \\
e^- e^+ &\rightarrow Z\gamma \\
e^- e^+ &\rightarrow Z\eta \\
\nu\bar{\nu} &\rightarrow Z Z \\
Z Z &\rightarrow Z Z
\end{aligned}
\tag{5.97}
$$

Of course, the processes (5.97) are most intriguing in the case of longitudinally polarized external vector bosons (note that we have already investigated the reaction $\bar{\nu} e \rightarrow W^- \gamma$ in Chapter 4). One may show, by means of an explicit calculation, that in tree-level scattering amplitudes of processes (5.97) the high-energy divergences cancel automatically, owing to the structure of the interaction Lagrangian (5.95). The corresponding calculations are left to the interested reader as an instructive exercise (see the problem 5.12). Besides that, the possible remaining non-leading divergences in scattering amplitudes of some processes (5.96) which we have considered earlier (such as $W^-W^+ \rightarrow ZZ$ or $W^-W^+ \rightarrow Z\gamma$) can be shown to vanish as well. The above-mentioned automatic divergence cancellations in tree-level amplitudes of the processes (5.97) etc. represent a fact remarkable in itself — they indicate that the Lagrangian (5.95) is at least a viable candidate for a reasonable theory of electroweak interactions. However, it is not at all clear whether the corresponding cancellations of unwanted terms for $E \rightarrow \infty$ occur in scattering amplitudes of *all* physical processes. In other words, the following two questions arise naturally:

1. Does the model (5.95) satisfy the condition of tree unitarity?

2. Does the model (5.95) satisfy the stronger condition (5.4), i.e. is the corresponding perturbation expansion renormalizable?

The answer to the first question is *yes* while the second question is to be answered in the *negative*. This statement (which we have already foreshadowed at the end of Section 5.1 but still may sound somewhat surprising) deserves a more detailed commentary. In the first place, one has to note that for technical reasons it is virtually impossible to verify directly the validity of the tree-unitarity condition for all (n-point) scattering amplitudes by means of the elementary methods employed so far. Fortunately one may proceed in a completely different manner. The interaction Lagrangian (5.95), which we have deduced through a systematic elimination of high-energy divergences in some selected tree-level Feynman diagrams, *is in fact identical with the original Weinberg model* [7] *of the unification of weak and electromagnetic interactions of leptons*. Of course, the Weinberg model [7] has been formulated as a non-abelian gauge theory with the Higgs mechanism (the vector bosons $W^\pm$, Z and γ correspond to the four gauge fields of the group $SU(2) \times U(1)$ and η is the Higgs boson; the Lagrangian (5.95) represents the particular choice of gauge condition used originally by Weinberg [7] - the so-called unitary, or U-gauge, which is characterized by absence of unphysical fields). For a detailed investigation of properties of a theory described by the Lagrangian (5.95) one may therefore employ the powerful formal apparatus of gauge theories (see e.g. [15], [17], [25] etc.). The complete tree-level unitarity in a theory of such a type was first proved by J. S. Bell who followed an earlier work of S. Weinberg (see [44]).

As to the stronger condition (5.4), it is violated at the level of one-loop diagrams. That is to say, one may find an example of a binary process, for which the corresponding scattering amplitude in the one-loop approximation (more precisely, its real part) grows linearly with energy, although imaginary parts of the relevant graphs (which of course are fully determined by the corresponding tree-level amplitudes) are asymptotically constant for $E \rightarrow \infty$ (cf. the discussion related to Fig. 11 in Section 5.1). The reason for such a "pathological" behaviour is the famous Adler-Bell-Jackiw (ABJ) axial anomaly in a triangular closed fermionic loop (the fermions are leptons in our case) [40], [45], [46]. This remarkable phenomenon will be discussed in more detail in the next section. Here we restrict ourselves to the following three closing remarks.

i) The above-mentioned linear growth of some one-loop scattering amplitudes for $E \to \infty$ implies non-renormalizability of the perturbation expansion on the level of two-loop diagrams.

ii) The effect of the ABJ anomaly demonstrates that *the tree unitarity is only a necessary, but not sufficient, condition for perturbative renormalizability* (as we have already indicated in Section 5.1).

iii) Despite the fact that subtle effects of the ABJ anomaly violate perturbative renormalizability of the theory described by the Lagrangian (5.95), in fact we have almost reached our objective (as (5.95) represents precisely the original Weinberg model [7]); in the following sections 5.6 and 5.7 we shall see that effects of the anomaly are removed "miraculously" (and at the same time very naturally from the physical point of view) if one considers the corresponding interactions of quarks, as well as electroweak interactions of leptons [45], [46] (in building a realistic theory of electroweak unification we are of course obliged to include the quark sector, in view of the phenomenologically well-established form of the weak charged current in (1.1) – (1.4)).

5.6 Effects of the ABJ Axial Anomaly

To illustrate a violation of the condition of "perturbative unitarity" (5.4) at the level of one-loop Feynman graphs, we shall consider, as an example, the process $e^+e^- \to \gamma\gamma$ (which is completely innocuous at the tree level). The diagrams leading to an "anomalous" behaviour of the corresponding one-loop scattering amplitude in the high-energy limit (in the sense indicated at the end of the preceding section) are shown in Fig. 30. Before examining these graphs in more detail, the following remark is in order: Within the framework of the theory described by the interaction Lagrangian (5.95), there are of course many other one-loop graphs besides those depicted in Fig. 30, contributing to the scattering amplitude of the considered process, but all of them already exhibit a "normal" behaviour in the high-energy limit (i.e. obey the law (5.4)). This fact can be best explained using the formalism of non-abelian gauge theories with the Higgs mechanism and therefore we will refrain from discussing it here.

We will now examine in more detail the contribution of the diagrams in Fig. 30(a), (b). The corresponding scattering amplitude may be written as

$$
\mathcal{M}_\Delta = \mathcal{M}_a + \mathcal{M}_b =
$$
$$
= i \left(\frac{g}{\cos \vartheta_W} \right)^2 a_e Q_e^2 e^2 \bar{v}(l_+) \gamma_\lambda (v_e - a_e \gamma_5) u(l_-) \times
$$
$$
\times \frac{-g^{\lambda\alpha} + m_Z^{-2} q^\lambda q^\alpha}{q^2 - m_Z^2} T_{\alpha\mu\nu}(k, p) \varepsilon^{*\mu}(k) \varepsilon^{*\nu}(p), \qquad (5.98)
$$

where the notation employed in (5.98) has the following meaning: Coupling constants for the interaction of weak neutral current with the Z correspond to formulae (5.37); here we have only introduced extra symbols for a vector and axial-vector interaction constant (indicating explicitly the lepton type)

$$\frac{1}{2}(g_L + g_R) = \frac{g}{\cos \vartheta_W} v_e$$

$$\frac{1}{2}(g_L - g_R) = \frac{g}{\cos \vartheta_W} a_e \qquad (5.99)$$

i.e. (see (5.37))

$$v_e = -\frac{1}{4} + \sin^2 \vartheta_W$$

$$a_e = -\frac{1}{4} \qquad (5.100)$$

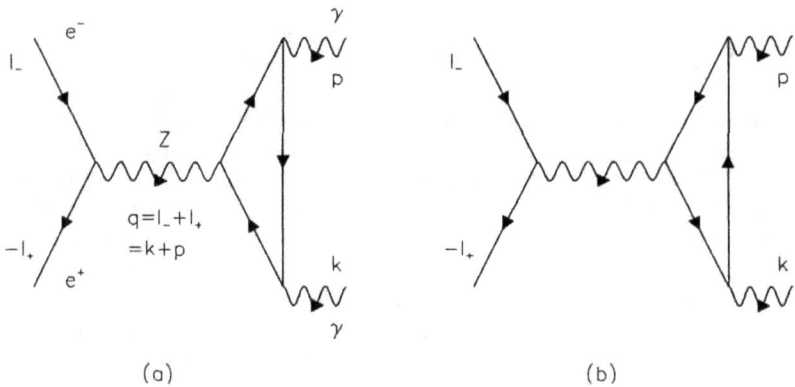

Fig. 30. *The one-loop diagrams of the process $e^+e^- \rightarrow \gamma\gamma$ in which an effect of the ABJ axial anomaly is manifested. Internal lines in the closed fermion loop correspond to a lepton (e.g. electron).*

In the expression (5.98) the coupling parameter a_e is factorized, since in the vertex of the triangular fermion loop attached to the Z propagator only the axial-vector part of the corresponding neutral current can play a role; the contribution of the vector part vanishes identically as a consequence of the well-known identities for traces of Dirac γ-matrices (the so-called Furry theorem - see e.g. [20], [21]). Further, for each electromagnetic interaction we have singled out explicitly a factor Q_e ($Q_e = -1$), i.e. the charge of the fermion in the closed loop in units of e (this is useful with regard to a later discussion of the quark sector). Finally, the expression $T_{\alpha\mu\nu}(k,p)$ is the contribution of the closed loops in Fig. 30(a), (b) which is given formally by the integral

$$
\begin{aligned}
T_{\alpha\mu\nu}(k,p) &= \int \frac{d^4 r}{(2\pi)^4} \mathrm{Tr}\left(\frac{1}{\not{r} - \not{k} - m}\gamma_\mu \frac{1}{\not{r} - m}\gamma_\nu \frac{1}{\not{r} + \not{p} - m}\gamma_\alpha\gamma_5 \right) \\
&+ [(k,\mu) \leftrightarrow (p,\nu)]
\end{aligned}
\tag{5.101}
$$

(it is easy to verify that the reversed orientation of the closed loop in the diagram (b) with respect to (a) corresponds to the symmetrization incorporated in Eq. (5.101)).

The integral in Eq. (5.101) is apparently (linearly) divergent in the ultraviolet region and thus it is an ill-defined object by itself; one should therefore add to the formal expression (5.101) a prescription giving it a precise meaning. It is well known (see [40], [47], [48]) that one can do so either by an appropriate regularization procedure or by imposing a physical requirement of absence of "longitudinally polarized photons" in the final state, i.e. by imposing the identities

$$
k^\mu T_{\alpha\mu\nu}(k,p) = 0, \qquad p^\nu T_{\alpha\mu\nu}(k,p) = 0
\tag{5.102}
$$

In the standard language of quantum field theory the relations (5.102) are usually called in this context "vector Ward identities" and they also express conservation of vector (electromagnetic) currents in the corresponding vertices of the considered Feynman graph. The construction of a finite quantity $T_{\alpha\mu\nu}$ based on the constraints (5.102) was first performed by Rosenberg [49] and employed later by Adler [48] in his pioneering investigation of the triangle anomaly. However, a more detailed discussion of various definitions of the $T_{\alpha\mu\nu}$ goes beyond the framework of this introductory treatment of the electroweak unification; in addition to the literature we have already quoted, one may find an elementary introduction to the anomaly problem in the textbooks [21], [25], [36] and also in the review article [50].

As regards the high-energy behaviour of the amplitude (5.98), its potentially "dangerous" part obviously corresponds to the second term in the Z propagator (because of presence of the factor m_Z^{-2}). Using the q^λ from this term to multiply the γ_λ in the first neutral-current vertex, the electron mass m is factorized (through an application of the Dirac equation), which compensates one factor of m_Z^{-1}. Multiplying the axial-vector vertex of the triangular fermion loop by the q^α, a naive calculation (in a sense described in detail e.g. in [50]) would lead to the conclusion that the expression $q^\alpha T_{\alpha\mu\nu}$ is equal to $2mT_{\mu\nu}$, where $T_{\mu\nu}$ is the contribution of the corresponding fermionic loops in which $\gamma_\alpha\gamma_5$ is replaced by γ_5 (such a result would correspond to a classic relation for the divergence of the axial-vector current). However, in fact (using a mathematically correct calculation procedure), the amplitude $T_{\alpha\mu\nu}$ subject to constraints (5.102) satisfies an axial-vector Ward identity

$$
q^\alpha T_{\alpha\mu\nu}(k,p) = 2mT_{\mu\nu}(k,p) + \frac{1}{2\pi^2}\varepsilon_{\mu\nu\rho\sigma}k^\rho p^\sigma,
\tag{5.103}
$$

where the second term on the right-hand side of eq. (5.103) is just the celebrated Adler-Bell-Jackiw (ABJ) axial anomaly. Since the fermion mass is not factorized in this anomalous term (a factor of m only appears in the first term on the right-hand side of (5.103)), there remains an uncompensated factor m_Z^{-1} in the contribution of Fig. 30(a), (b) and the corresponding amplitude thus grows linearly with energy for $E \to \infty$.

It should be emphasized (as indicated in the preceding section) that the *imaginary* (or "absorptive") part of the contribution of diagrams in Fig. 30(a), (b) is *finite* in the limit $E \to \infty$. (The following technical remark is in order here: The terms "imaginary" or "absorptive" part are commonly used in an equivalent sense; a non-zero absorptive part corresponds to a discontinuity on a cut which in the considered case exists on the real axis of the variable $s = q^2$ for $s > 4m^2$. For a general discussion of such singularities and analytic properties of scattering amplitudes and Green functions in quantum field theory see e.g. in [21] where one may also find a formulation of the standard Cutkosky rules for computing the absorptive part of a Feynman graph.) The finiteness of the absorptive part of the diagrams in Fig. 30 in the high-energy limit may easily be understood if one realizes that this can be expressed by means of a product of amplitudes of tree diagrams corresponding to processes $e^+e^- \to e^+e^-$ and $e^+e^- \to \gamma\gamma$ (see Fig. 31(a), (b)). These tree-level graphs are of course finite in the limit $E \to \infty$: In the case of the diagram for $e^+e^- \to e^+e^-$ involving the Z exchange, a factor m^2 is produced in the potentially offending term (when the Dirac equation is applied in both vertices) which compensates the m_Z^{-2} from the Z propagator; the "good" behaviour of the tree-level graph for $e^+e^- \to \gamma\gamma$ is manifest. In calculating the contribution of Fig. 31(a), (b) one must of course also integrate over the phase-space volume for the electron-positron intermediate states; such an integration, however, does not change qualitatively the estimate inferred from the behaviour of tree-level graphs of the intermediate processes. It is instructive to demonstrate the difference between asymptotic behaviour of the diagrams in Fig. 30 and of their absorptive part (Fig. 31) in terms of explicit formulae. For the amplitude of the considered fermion triangular loops $T_{\alpha\mu\nu}$ (see Eq. (5.101)) satisfying the conditions (5.102) one may write a tensor decomposition (for a detailed discussion see e.g. [50 – 52])

$$T_{\alpha\mu\nu}(k,p) = F_1(s)q_\alpha\varepsilon_{\mu\nu\rho\sigma}k^\rho p^\sigma +$$
$$+ F_2(s)(p_\nu\varepsilon_{\alpha\mu\rho\sigma} - k_\mu\varepsilon_{\alpha\nu\rho\sigma})k^\rho p^\sigma \qquad (5.104)$$

where we have used the notation $s = q^2$; for the validity of Eq. (5.104) it is essential that $k^2 = p^2 = 0$. The invariant amplitudes (formfactors) F_1 and F_2 may be expressed as integrals over Feynman parameters

$$F_1(s) = -\frac{1}{\pi^2} \int_0^1 dx \int_0^{1-x} dy \frac{xy}{m^2 - xys - i\varepsilon} \qquad (5.105)$$

$$F_2(s) = \frac{1}{\pi^2} \int_0^1 dx \int_0^{1-x} dy \frac{x(1 - x - y)}{m^2 - xys - i\varepsilon} \qquad (5.106)$$

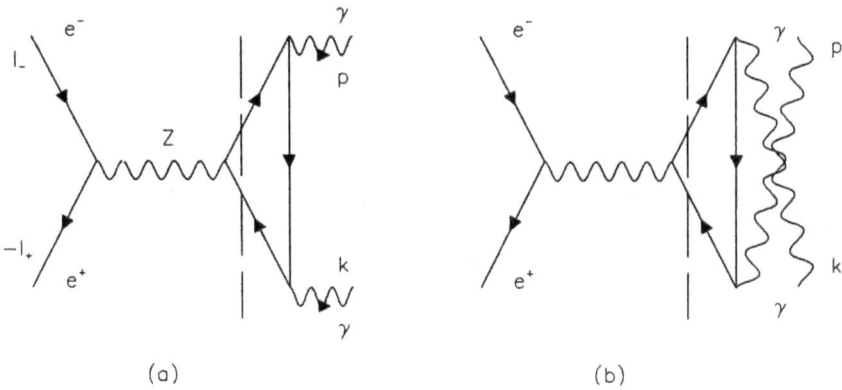

Fig. 31. *Absorptive part of the diagrams from Fig. 30. The permutation of the external photon lines in the graph (b) is equivalent to reversing the orientation of the closed fermion loop in Fig. 30(b). The usual notation is used, such that the "cut" internal lines correspond to real particles, i.e. the corresponding propagators are replaced by δ-functions according to the Cutkosky rules.*

In the first place one may now verify the anomalous Ward identity (5.103): From Eq. (5.104) it follows immediately that

$$q^\alpha T_{\alpha\mu\nu} = s F_1(s) \varepsilon_{\mu\nu\rho\sigma} k^\rho p^\sigma$$

and from (5.105) one gets

$$s F_1(s) = -\frac{1}{\pi^2} \int_0^1 dx \int_0^{1-x} dy \frac{m^2}{m^2 - xys - i\varepsilon} + \frac{1}{2\pi^2} \tag{5.107}$$

where the first expression on the right-hand side of Eq. (5.107) corresponds to the "normal" term in (5.103) (which vanishes for $m \to 0$) and the second term reproduces the ABJ anomaly. On the basis of Eq. (5.107) one may also easily estimate the asymptotic behaviour of the function $F_1(s)$ in the limit $s \to \infty$ (i.e. pro $s \gg m^2$):

$$F_1(s) = \frac{1}{2\pi^2} \frac{1}{s} + O\left(\frac{m^2}{s^2} \ln \frac{s}{m^2}\right) \tag{5.108}$$

The absorptive (i.e. imaginary) part of the amplitude $T_{\alpha\mu\nu}$ which we denote as $A_{\alpha\mu\nu}$ may be written in the form analogous to (5.104) where the formfactors F_1 and F_2 are replaced by the corresponding imaginary parts $A_j = \operatorname{Im} F_j$, $j = 1, 2$. A calculation of the A_1 and A_2 using the formulae (5.105), (5.106) is straightforward and yields the results

$$A_1(s) = -\frac{1}{\pi}\frac{m^2}{s^2}\ln\frac{1 + \sqrt{1 - 4m^2/s}}{1 - \sqrt{1 - 4m^2/s}} \tag{5.109}$$

$$A_2(s) = \frac{1}{2\pi}\left(\frac{\sqrt{1 - 4m^2/s}}{s} - \frac{2m^2}{s^2}\ln\frac{1 + \sqrt{1 - 4m^2/s}}{1 - \sqrt{1 - 4m^2/s}}\right) \tag{5.110}$$

(the formula (5.109) was first used in connection with the ABJ axial anomaly by Dolgov and Zakharov [53]). From (5.109), (5.110) one easily gets the corresponding asymptotic expressions valid for $s \to \infty$ (i.e. for $s \gg m^2$):

$$A_1(s) = O\left(\frac{m^2}{s^2}\ln\frac{s}{m^2}\right) \tag{5.111}$$

$$A_2(s) = \frac{1}{2\pi}\frac{1}{s} + O\left(\frac{m^2}{s^2}\ln\frac{s}{m^2}\right) \tag{5.112}$$

From Eq. (5.108) and (5.111) it is obvious that the high-energy behaviour of the real part of the formfactor F_1 differs substantially from the asymptotics of the corresponding imaginary part: While the imaginary part decreases for $s \to \infty$ as $1/s^2$ (up to a logarithmic factor) the real part only falls off as $1/s$.

The following technical remark is in order here. Formulae (5.105), (5.106) are obtained by a direct computation of the amplitude $T_{\alpha\mu\nu}$ and from these one may derive the expressions (5.109), (5.110) for the corresponding imaginary part $A_{\alpha\mu\nu}$. However, one may also proceed in a reversed order: Using the well-known Cutkosky rules (see e.g. [21]) one may first calculate the absorptive part $A_{\alpha\mu\nu}$ (let us stress that this is given by convergent integrals) and the full formfactors F_1, F_2 may then be defined by means of dispersion relations (which in the considered case converge without subtractions). The above-mentioned difference in the power behaviour of the $A_1(s)$ and $F_1(s)$ for $s \to \infty$ can then be traced, technically, to the integration in the corresponding dispersion relation

$$F_1(s) = \frac{1}{\pi}\int_{4m^2}^{\infty}\frac{A_1(s')}{s' - s}ds' $$

(this is precisely the effect we have mentioned in a preliminary discussion in the introductory Section 5.1). However, as we have noticed earlier, a fundamental reason for this effect is the ABJ axial anomaly; within the framework of the dispersion relation approach (which in this case obviates completely the problem of ultraviolet divergences) the anomaly is a consequence of special properties of the invariant amplitude A_1, in particular, of the fact that the integral of the A_1 taken along the cut $(4m^2, \infty)$ is non-zero. Indeed, for the function $A_1(s; m^2)$ given by the formula (5.109) one has (for an arbitrary value of m) a "sum rule"

$$\int_{4m^2}^{\infty}A_1(s; m^2) = -\frac{1}{2\pi}. \tag{5.113}$$

(It is interesting to notice that a dominant contribution to the integral (5.113) comes from the region of small s, i.e. from the vicinity of the threshold $s_0 = 4m^2$. As we have already indicated, such an interpretation of the ABJ anomaly was first formulated in the paper [53]; a brief review of the method as well as further details can also be found e.g. in [51], [52].)

After this rather technical exposition we are going to discuss again the part of the contribution of diagrams in Fig. 30 or Fig. 31, which corresponds to the longitudinal term in the Z propagator. From what we have already said it may easily be seen that the considered part of the scattering amplitude (which we will denote as $\mathcal{M}_\Delta^{(2)}$) behaves in the high-energy limit (i.e. for $s \gg m_Z^2$) as

$$\mathcal{M}_\Delta^{(2)} \simeq \bar{v}(l_+)\gamma_5 u(l_-)\frac{m}{m_Z^2}\frac{1}{s}\varepsilon_{\mu\nu\rho\sigma}k^\rho p^\sigma \varepsilon^{*\mu}(k)\varepsilon^{*\nu}(p) \tag{5.114}$$

(i.e. it grows linearly for $E \to \infty$ as we have already stated). For the absorptive part of the $\mathcal{M}_\Delta^{(2)}$ we obtain an asymptotic estimate

$$\text{Abs } \mathcal{M}_\Delta^{(2)} \simeq \bar{v}(l_+)\gamma_5 u(l_-)\frac{m}{m_Z^2}\frac{m^2}{s^2}\ln\frac{s}{m^2}\varepsilon_{\mu\nu\rho\sigma}k^\rho p^\sigma \varepsilon^{*\mu}(k)\varepsilon^{*\nu}(p) \tag{5.115}$$

i.e. *Abs* $\mathcal{M}_\Delta^{(2)}$ falls off as $1/E$ for $E \to \infty$.

As regards the diagonal term in the Z propagator and the corresponding part of the contribution of diagrams in Fig. 30 (we shall denote this part by $\mathcal{M}_\Delta^{(1)}$) one may easily estimate on the basis of the above-mentioned relations that in the limit $E \to \infty$

$$\mathcal{M}_\Delta^{(1)} \simeq O(1) \tag{5.116}$$

and also

$$\text{Abs } \mathcal{M}_\Delta^{(1)} \simeq O(1) \tag{5.117}$$

(the last estimate follows from the fact that *Abs* $\mathcal{M}_\Delta^{(1)}$ gets a contribution from the invariant amplitude A_2 which according to (5.12) decreases for $s \to \infty$ as $1/s$).

Let us supplement the preceding discussion with the following remark: A preliminary semi-quantitative estimate of the asymptotic behaviour of the *Abs* $\mathcal{M}_\Delta$, based on considerations about tree-level graphs of the intermediate processes in Fig. 31 which we have formulated earlier in this section, can be made more precise using a formula for the absorptive part of the relevant triangle diagram (cf. e.g. [54])

$$2i\, A_{\alpha\mu\nu}(k,p) = -\frac{1}{32\pi^2}\frac{|\vec{P}|}{E}\sum_{s,s'}\int d\Omega[\bar{u}(P,s)\gamma_\alpha\gamma_5 v(P',s')] \times$$
$$\times\; [\bar{v}(P',s')\gamma_\nu\frac{P-\not{k}+m}{(P-k)^2-m^2}\gamma_\mu u(P,s)], \tag{5.118}$$

which may by derived either directly from S-matrix unitarity or using Cutkosky rules. The relation (5.118) is written in the c.m. system of the pair of photons in the final state. The four-momenta P, P' are of course on the mass shell, i.e. they satisfy conditions $P^2 = P'^2 = m^2$; one has further $(P + P')^2 = (k + p)^2 = s$ and $\vec{P} = -\vec{P}'$, so $P_0 = P'_0 = E = \frac{1}{2}\sqrt{s}$, $|\vec{P}| = \frac{1}{2}\sqrt{s - 4m^2}$. The factor $|\vec{P}|/E = \sqrt{1 - 4m^2/s}$ in (5.118) comes from the phase-space volume of the two-particle intermediate state e^+e^- and the angular integration is carried out over directions of the $\vec{P}$. From (5.118) one then immediately gets an appropriate relation for *Abs* $\mathcal{M}_\triangle$ which enables one to verify the corresponding statements made earlier.

Let us emphasize the main result of this section, namely the observation of a linear growth of the considered amplitude $\mathcal{M}_\triangle$ with energy in the limit $E \to \infty$. It is important to realize that (if we consider only a dependence on properties of the fermion in the anomalous triangular loop in Fig. 30) the relevant numerical coefficient in the corresponding leading asymptotic term is

$$C^{(e)}_{anomaly} = a_e Q_e^2 \qquad (5.119)$$

The origin of (5.119) is obvious from the discussion around (5.98) and (5.103). From (5.119) it is obvious that the linear divergence of the amplitude $\mathcal{M}_\triangle$ for $E \to \infty$ cannot be compensated or removed if we take into account only the electroweak interactions of leptons; a neutrino loop of course does not contribute to the considered process and all the standard charged leptons (as e.g. muon) give a contribution identical with (5.119), since for an arbitrary charged lepton l one may obviously repeat the procedure described in Section 5.3 and thus arrive at the result

$$a_l = -\frac{1}{4} \qquad (5.120)$$

(of course, one always has $Q_l^2 = 1$). The condition (5.4) is thus violated and the model described by the interaction Lagrangian (5.95), which incorporates only leptons in its fermionic sector, is therefore not renormalizable. Let us remark that (as indicated at the end of the preceding section) non-renormalizable ultraviolet divergences will appear in diagrams involving at least two closed loops (this of course is closely related to the power-like growth of the corresponding anomalous one-loop graphs for $E \to \infty$). An explicit example of a 2-loop graph leading to a non-renormalizable ultraviolet divergence is given in Fig. 32 (for a more detailed discussion see e.g. [46]). In the following section we will show, among other things, that the contribution of anomalous triangular loops made of quark fields can cancel the lepton contribution completely (see the original papers [45], [46] and also e.g. [21] and [25]).

As a conclusion to this section let us add that within the framework of gauge theories with Higgs mechanism one encounters other possible manifestations of the ABJ anomaly, as e.g. a gauge-dependence of physical scattering amplitudes (at the one-loop level) or a violation of unitarity of the S-matrix (at the two-loop level); for a more detailed discussion of these effects, see e.g. [46] and also the textbook [17]. However, it should be stressed again that all the "destructive" effects of the ABJ anomaly manifested in the perturbation expansion in fact disappear when both leptons and quarks are incorporated in the fermion sector of the standard model of electroweak interactions and the resulting theory is then perturbatively renormalizable.

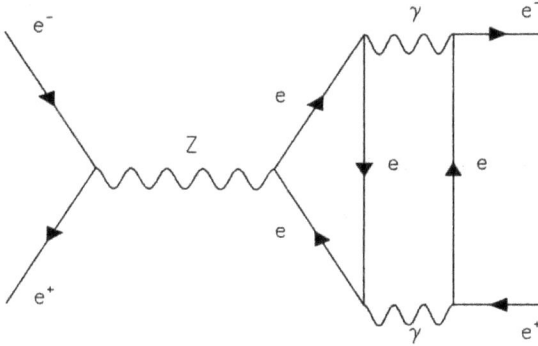

Fig. 32. *An example of a 2-loop diagram of the process* $e^-e^+ \rightarrow e^-e^+$ *in which the ABJ anomaly induces a non-renormalizable ultraviolet divergence.*

5.7 Interactions in the Quark Sector

Now we will investigate weak and electromagnetic interactions of hadrons; it is natural (and physically well substantiated) to describe these as interactions of fundamental quark fields, i.e. the fields of elementary fermions with fractional electric charges (fractional with respect to the charge of electron or muon). In Chapter 1 we gave a form of the weak charged quark current (i.e. the current constructed from fields carrying charges differing by one unit) expressed in terms of fields of the four quarks u, d, s, c (see (1.4)). (Needless to say, u = "up", d = "down", s = "strange", c = "charm".) The starting point of our discussion in this section will be the original weak current of the Cabibbo type, corresponding to the first line in the expression (1.4) (which involves only u, d and s). That is, we will not assume *a priori* the existence of a c-quark (which indeed was confirmed only after a theoretical prediction) and we will show that one may arrive naturally at the concept of an extra quark field through considerations concerning the high-energy behaviour of some tree-level diagrams, supplemented with some well-known facts about phenomenology of weak processes. In other words, within our general approach based on an investigation of tree-level amplitudes of elementary binary processes we will *derive* the structure of the weak charged current involving the c-quark (i.e. the second line in (1.4)).

The expression (1.4) corresponds to a realization of the familiar Glashow-Iliopoulos-Maiani (GIM) mechanism [55] (see also [25], [56], [57]), i.e. to a suppression of weak *neutral* currents non-diagonal with respect to "flavours" of the type u, d, s or c (let us stress that the weak *charged* current (1.4) is *non-diagonal*). The meaning of such a mechanism will be clarified in the subsequent discussion.

Let us first consider the interaction of the quark current of the Cabibbo type with the field of charged intermediate vector bosons described by the Lagrangian

$$\mathcal{L}_{CC}^{(u,d,s)} = \frac{g}{2\sqrt{2}}\bar{u}\gamma_\rho(1 - \gamma_5)(d\cos\vartheta_C + s\sin\vartheta_C)W^{+\rho} + \text{h.c.} \qquad (5.121)$$

In the tree approximation we shall examine the scattering amplitude of the process

$$d\bar{s} \to W^-W^+ \qquad (5.122)$$

Within the model described by the interaction Lagrangian (5.121) there is a single diagram corresponding to the process (5.122), namely that depicted in Fig. 33 (in this case the electromagnetic interaction does not contribute, as the electromagnetic current is flavour-diagonal). We are going to discuss the high-energy behaviour of the tree-level amplitude of the process (5.122) when both final-state vector bosons have longitudinal polarizations. Using the by-now-familiar arguments one may guess immediately that the contribution of the diagram in Fig. 33 diverges quadratically for $E \to \infty$. Let us denote the corresponding scattering amplitude by $\mathcal{M}^{(u)}$ (to indicate the u-quark exchange in Fig. 33); an explicit calculation (which is completely analogous to procedures used earlier in the lepton sector) yields the result

$$
\begin{aligned}
\mathcal{M}^{(u)} &= -\frac{1}{4m_W^2}g_{ud}g_{us}\bar{v}(l)\,\not{p}\,(1 - \gamma_5)u(k) \\
&\quad - m_s\frac{1}{4m_W^2}g_{ud}g_{us}\,\bar{v}\,(l)(1 - \gamma_5)u(k) \\
&\quad + O(1)
\end{aligned}
\qquad (5.123)
$$

(see the problem 5.13). In Eq. (5.123) we have introduced a natural notation (cf. (5.121))

$$g_{ud} = g\cos\vartheta_C, \qquad g_{us} = g\sin\vartheta_C \qquad (5.124)$$

The first term on the left-hand side of Eq. (5.123) represents the leading (quadratic) divergence and the second term corresponds to a next-to-leading (linear) divergence in the limit $E \to \infty$. (It is important to notice that none of the divergent terms in (5.123) depend on m_u, i.e. even the non-leading divergence is independent of the mass of the exchanged u quark; this circumstance will play an essential role in the divergence cancellation mechanism working in the high-energy limit for the considered scattering amplitude.)

We might attempt to cancel the quadratic divergence in (5.123) (in analogy with the case of the process $e^+e^- \to W^+W^-$ etc.) by means of a tree diagram involving an

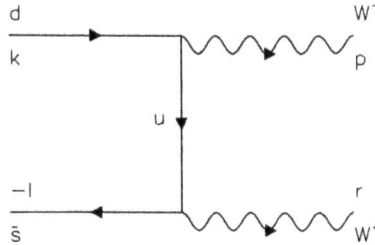

Fig. 33. *Tree-level diagram of the process* $d\bar{s} \to W^-W^+$ *in a model of weak interactions involving a quark charged current of the Cabibbo type.*

exchange of the neutral vector boson Z in the s-channel; to this end one would have to introduce a direct interaction of the type $\mathcal{L}_{dsZ}$, i.e. an interaction of the Z with a weak neutral strangeness-changing current. The corresponding coupling constant would then have to be of the order of

$$g_{dsZ} \simeq g \sin \vartheta_C \cos \vartheta_C / \cos \vartheta_W \qquad (5.125)$$

(cf. (5.124) and (5.37)). However, the existence of such an interaction would lead to a phenomenological disaster, in the sense that it would be clearly incompatible with common experimental data: Strangeness-changing ($\Delta S \neq 0$) decay processes in which the hadron charge is conserved ($\Delta Q = 0$) would be predicted within such a theory to occur in the lowest perturbative order, so the corresponding decay rates would have to be comparable with those of the commonly observed processes for which $\Delta Q \neq 0$, $\Delta S \neq 0$ (let us recall that the allowed decays obey the empirical selection rule $\Delta S = \Delta Q$). In fact, the data show that the weak processes in which $\Delta Q = 0$ and $\Delta S \neq 0$ are strongly suppressed in comparison with the cases $\Delta S \neq 0$, $\Delta Q \neq 0$. Thus, e.g. the relative decay rate (branching ratio) of the process $K^- \to \pi^0 e^- \bar{\nu}_e$ (i.e. $s \to u e^- \bar{\nu}_e$ on the quark level) is

$$\mathrm{BR}\,(K^- \to \pi^0 e^- \bar{\nu}_e) \doteq 0.048$$

(see [58]) whereas in the case of the decay $K^- \to \pi^- e^+ e^-$ (which corresponds to $s \to d e^+ e^-$ on the quark level) one has [58]

$$\mathrm{BR}\,(K^- \to \pi^- e^+ e^-) \doteq 2.7 \times 10^{-7}$$

There are other examples of such a type, so one may conclude that introducing a direct interaction of the Z with a strangeness-changing neutral current is phenomenologically unacceptable.

As regards the other conceivable mechanisms for suppression of power-like high-energy divergences in (5.123) (within the general scheme delineated in Section 5.2) it is also clear that an exchange of a scalar particle is not sufficient for the compensation of the quadratic divergence (cf. the discussion around the relations (5.6) - (5.8)) and thus we are left with the last possibility: One may attempt to cancel the offending terms in (5.123) by adding to the diagram in Fig. 33 a similar one, in which instead of the u-quark exchange another spin-$\frac{1}{2}$ fermion is involved. For this purpose (and within our "minimal strategy") we are going to introduce another quark (denoted as c) with the same charge as the u (i.e. $Q_c = Q_u = 2/3$) and the corresponding interaction with d, s and with vector bosons $W^\pm$ will be assumed to have a form analogous to (5.121), i.e.

$$\mathcal{L}_{CC}^{(c,d,s)} = \frac{1}{2\sqrt{2}}\Big[\bar{c}\gamma_\rho(1-\gamma_5)(g_{cd}d + g_{cs}s)\Big]W^{+\rho} + \text{h.c.} \tag{5.126}$$

(cf. the notation (5.124)) where g_{cd} and g_{cs} are the corresponding (in general complex) coupling constants which must be determined. The tree diagram for the process $d\bar{s} \to W^-W^+$ corresponding to the interaction (5.126) is shown in Fig. 34. For its contribution (which we denote as $\mathcal{M}^{(c)}$) for longitudinally polarized $W^\pm$ and using Eq. (5.123) we immediately get

$$\begin{aligned}
\mathcal{M}^{(c)} &= -\frac{1}{4m_W^2}g_{cd}g_{cs}^*\bar{v}(l)\,\slashed{p}\,(1-\gamma_5)u(k) \\
&\quad - m_s\frac{1}{4m_W^2}g_{cd}g_{cs}^*\,\bar{v}\,(l)(1-\gamma_5)u(k) \\
&\quad + O(1)
\end{aligned} \tag{5.127}$$

Quadratic divergences in Eq. (5.123) and (5.137) cancel each other if and only if

$$g_{ud}g_{us} + g_{cd}g_{cs}^* = 0 \tag{5.128}$$

It is gratifying that the condition (5.128) automatically guarantees even a cancellation of linear divergences in Eq. (5.123) and (5.129); so we need not worry about any extra strangeness-changing neutral scalar exchange (which would be phenomenologically unacceptable). Of course, the observed automatic cancellation of linear divergences is due to the fact (which we have emphasized earlier) that these terms do not depend on the mass of the exchanged quark in diagrams in Fig. 33, 34.

The relation (5.128) gives one constraint for two unknown coupling constants g_{cd}, g_{cs}. However, now one may also consider the process

$$u\bar{c} \to W^-W^+ \tag{5.129}$$

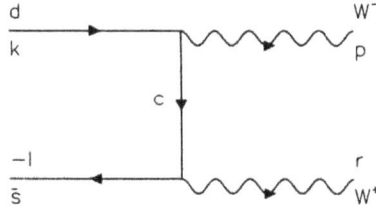

Fig. 34. *The diagram of the process* $d\bar{s} \to W^-W^+$ *involving an exchange of the c-quark which compensates the divergent behaviour of Fig. 33 in the high-energy limit.*

which in the lowest order in (5.121), (5.126) proceeds via the diagrams shown in Fig. 35. As before, we are going to discuss the case of longitudinally polarized W's. Following essentially the same steps which have previously led to (5.128) one finds that the cancellation of high-energy divergences in diagrams (a) and (b) in Fig. 35 is equivalent to

$$g_{ud}\, g_{cd} + g_{us}\, g_{cs} = 0 \qquad (5.130)$$

(similarly to the relation (5.128), the condition (5.130) guarantees an elimination of quadratic as well as linear divergences in the corresponding tree-level scattering amplitude). It should be emphasized that fulfillment of (5.130) is also important from a phenomenological point of view, since recent experimental data show that the existence of a direct interaction of the type $\mathcal{L}_{ucZ}$ (i.e. an interaction of the corresponding neutral current and the Z) is as implausible as the $\mathcal{L}_{sdZ}$ which we discussed earlier (see [58]).

Equations (5.128), (5.130) for the unknown coupling constants g_{cd} and g_{cs} can now be solved easily. After a simple manipulation one gets first

$$\begin{aligned} |g_{cs}| &= g_{ud} \\ |g_{cd}| &= g_{us} \end{aligned} \qquad (5.131)$$

and using (5.124) we may write

$$\begin{aligned} g_{cd} &= g \sin \vartheta_C \exp\left(i\delta_{cd}\right) \\ g_{cs} &= g \cos \vartheta_C \exp\left(i\delta_{cs}\right) \end{aligned} \qquad (5.132)$$

where the phases δ_{cd}, δ_{cs} are real numbers. Substituting (5.132) and (5.124) into eq.

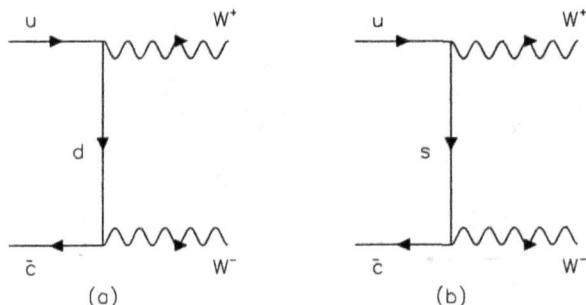

Fig. 35. *Tree-level diagrams of the process* $u\bar{c} \to W^- W^+$ *involving a d- and s-quark exchange.*

(5.130) we obtain

$$\exp\left(i\delta_{cd}\right) = -\exp(i\delta_{cs}) \tag{5.133}$$

The general solution of the system of equations (5.128), (5.130) is thus

$$\begin{aligned}
g_{cd} &= -g \sin \vartheta_C \exp\left(i\delta\right) \\
g_{cs} &= g \cos \vartheta_C \exp\left(i\delta\right)
\end{aligned} \tag{5.134}$$

where δ is an arbitrary real number.

If we now employ the result (5.134) in the interaction Lagrangian (5.126), it is easy to see that the phase δ is in fact irrelevant as it may be eliminated by means of a suitable redefinition of the c-quark field (in other words, the phase factor $\exp(i\delta)$ from the coupling constants may be "absorbed" in a definition of the dynamical variable of the c-quark field). The "compensation" Lagrangian (5.126) may thus, without loss of generality, be written as

$$\mathcal{L}_{CC}^{(c,d,s)} = \frac{g}{2\sqrt{2}}\bar{c}\gamma_\rho(1 - \gamma_5)(-d\sin \vartheta_C + s \cos \vartheta_C)W^{+\rho} + \text{h.c.} \tag{5.135}$$

The preceding considerations may be summarized as follows: Starting from a phenomenological model of weak interactions of the three quarks u, d, s involving non-trivial Cabibbo mixing (5.121), it is necessary to postulate the existence of another quark, if one wants to respect the tree unitarity and, at the same time, to avoid flavour-changing neutral currents. The requirement of tree unitarity also determines uniquely the structure of the relevant c-quark interaction (5.135): The corresponding charged current must contain a combination of the fields d and s which is "orthogonal" with respect to the original Cabibbo combination in (5.121).

The result (5.135) was first obtained by Glashow, Iliopoulos and Maiani [55] within the framework of gauge theory of weak and electromagnetic interactions based on the standard gauge group $SU(2) \times U(1)$. The suppression of unwanted effects of non-diagonal neutral currents, following from (5.135), is therefore called (as noted earlier in this section) the GIM mechanism. Introducing the c-quark is also very natural from an "aesthetic" point of view, more precisely from the point of view of a lepton-quark symmetry, since the four quarks u, d, s, c are then natural partners of the four leptons ν_e, e, ν_μ, μ. Such a symmetric scheme was in fact originally proposed by Bjorken and Glashow as early as 1964 [59] without anticipating its possible dynamic consequences. It should be stressed that the theoretical prediction of the c-quark [55], [59] has been remarkably successful since it has been experimentally confirmed (in a rather unexpected way) in 1974 as the "hidden charm" in the J/ψ particles (see [60]); a number of experiments performed in subsequent years then repeatedly demonstrated both the existence of charmed hadrons (i.e. the "overt charm") and various aspects of the GIM mechanism (see e.g. [25], [61], [62]). (In this context one should also recall that we did not have to discuss any analogy of the GIM mechanism in the lepton sector because we have not considered a priori any mixing between leptons of the electron and muon type; at present there is indeed no clear-cut experimental argument for introducing a phenomenological parameter analogous to the Cabibbo angle into leptonic weak interactions — see [58].)

The full Lagrangian describing weak interactions of the four quark fields u, d, s, c mediated by charged vector bosons may be denoted as $\mathcal{L}_{CC}^{(GIM)}$:

$$
\begin{aligned}
\mathcal{L}_{CC}^{(GIM)} &= \mathcal{L}_{CC}^{(u,d,s)} + \mathcal{L}_{CC}^{(c,d,s)} = \\
&= \frac{g}{2\sqrt{2}} \Big[\bar{u} \gamma_\rho (1 - \gamma_5)(d \cos \vartheta_C + s \sin \vartheta_C) \\
&\quad + \bar{c} \gamma_\rho (1 - \gamma_5)(-d \sin \vartheta_C + s \cos \vartheta_C) \Big] W^{+\rho} + \text{h.c.}
\end{aligned}
\tag{5.136}
$$

With regard to some future considerations it is convenient to recast the last expression in a matrix form as

$$
\mathcal{L}_{CC}^{(GIM)} = \frac{g}{2\sqrt{2}} (\bar{u}, \ \bar{c}) \gamma_\rho (1 - \gamma_5) V_{GIM} \begin{pmatrix} d \\ s \end{pmatrix} W^{+\rho} + \text{h.c.}
\tag{5.137}
$$

where V_{GIM} is the real orthogonal matrix

$$
V_{GIM} = \begin{pmatrix} \cos \vartheta_C & \sin \vartheta_C \\ -\sin \vartheta_C & \cos \vartheta_C \end{pmatrix}
\tag{5.138}
$$

For completeness we should now recall further well-known empirical facts about the spectrum of elementary fermions. In 1975 a new charged lepton, denoted as τ, with the rest mass of about $1.8 \text{ GeV}/c^2$, was discovered [63] (this of course does not coincide with a hypothetical "heavy lepton" mentioned in Section 5.2; the tau lepton is in a sense only a "copy" of the electron or muon and it carries a new conserved lepton charge). A corresponding neutrino ν has not been observed (in contrast to the ν_e or ν_μ) *directly* so far (i.e. the corresponding scattering experiments with the ν_τ have not yet been performed); however, in view of a lot of convincing indirect evidence, the existence of a ν_τ is generally assumed to be established (see [58]). Moreover, in 1977 there were published the first experimental data pointing toward the existence of another quark species, denoted as b ("bottom"), with charge $Q_b = -1/3$ (a brief review of the corresponding experimental results may be found e.g. in [61]). Assuming quite generally the above-mentioned lepton-quark symmetry, a natural counterpart of the six leptons (ν_e, e, ν_μ, μ, ν_τ, τ) should then be the same number of quarks; besides the experimentally established species ("flavours") u, d, s, c, b there should therefore exist another quark, commonly denoted as t ("top"), with the charge $Q_t = 2/3$. A direct evidence for the t-quark (i.e. an experimental detection of processes related to its existence) is generally expected to appear during the 1990's (a present experimental value of the corresponding "rest mass" is about $m_t = 174 \text{ GeV}/c^2$ [65]; for comparison, $m_b \doteq 5 \text{ GeV}/c^2$ and $m_c \doteq 1.5 \text{ GeV}/c^2$ see e.g. [61]). However, the reason for such an expectation is not only an "aesthetic" aspect of a quark-lepton symmetry. Indeed (as indicated at the end of Section 5.6), such a symmetry of the spectrum of elementary fermions within the framework of the standard model of electroweak interactions plays an important role in cancellation of the triangle ABJ anomalies; we will deal with this remarkable fact in more detail somewhat later. Moreover, there are compelling (though indirect) experimental arguments in favour of existence of the t-quark [68] (if we assume the validity of the basic principles of the theory of electroweak unification). Let us recall at least one of them: If we consider a model of the interaction of charged currents and vector bosons $W^\pm$ involving 3 quarks with charge equal to $-1/3$ (i.e. d, s, b) and only 2 quarks with charge $2/3$ (i.e. u, c) then in the presence of a non-trivial mixing among d, s, b (which is indeed confirmed by experiments — see [66], [67] and the review [68]) the condition of tree unitarity in annihilation channels with initial states $d\bar{s}$, $d\bar{b}$ and $s\bar{b}$ would force us to introduce the corresponding neutral currents and interactions of the type $\mathcal{L}_{dbZ}$ and $\mathcal{L}_{sbZ}$; the existence of such interactions is however unacceptable phenomenologically (see [69] and the review [68]). Flavour-changing neutral currents may be avoided if we assume the existence of a t-quark with specific properties; in this context, the role of the t-quark is analogous to that played by the c-quark in the GIM mechanism. A discussion of technical details of the indicated considerations is recommended to the reader as an instructive exercise (see the problem 5.14). The charged-current interactions in a model involving six quarks are then parametrized by means of elements of a unitary 3×3 matrix which is now usually called the Kobayashi-Maskawa, or Cabibbo-Kobayashi-Maskawa (CKM) matrix [70] (see also [58], [68]), and we thus get a generalization of the GIM interaction Lagrangian (5.137)

$$\mathcal{L}_{CC}^{(CKM)} = \frac{g}{2\sqrt{2}}(\bar{u}, \ \bar{c}, \ \bar{t})\gamma_\rho(1 - \gamma_5)V_{CKM}\begin{pmatrix} d \\ s \\ b \end{pmatrix} W^{+\rho} + \text{h.c.} \qquad (5.139)$$

where V_{CKM} is the above-mentioned unitary matrix

$$V_{CKM} = \begin{pmatrix} V_{ud} & V_{us} & V_{ub} \\ V_{cd} & V_{cs} & V_{cb} \\ V_{td} & V_{ts} & V_{tb} \end{pmatrix} \tag{5.140}$$

The matrix V_{CKM} can be described in terms of four physically relevant real parameters (if one employs a suitable redefinition of phases of the quark fields in (5.139)), viewed as three angles and one phase (which may be related to CP violation [70]). In [58] one may find a "standard" parametrization of such a type (see also [68] and the original papers [71]) as well as numerical values of matrix elements in (5.140). Methods of experimental determination of the matrix V_{CKM} (more precisely, its first two rows) are reviewed e.g. in [68]. One more terminological remark is in order here: In connection with the empirical structure of the spectrum of elementary fermions which is suggested by experiments (and which is also corroborated by the renormalizable theory of electroweak interactions), the notion of fermion "generations" has become customary in particle physics: Fermions of the first generation are ν_e, e, u, d, to the second generation belong ν_μ, μ, c, s and the third generation (incomplete as yet because of the missing top-quark) is defined to consists of ν_τ, τ, t and b.

Problems of the GIM mechanism and its generalization to a model involving six quarks (i.e. three generations of fermions) within the usual framework of non-abelian gauge theory with Higgs mechanism are treated in considerable detail e.g. in [25], [56], [57], [62] and [68]. For simplicity, in what follows we are going to discuss a model involving four quarks (i.e. two generations of fermions); a generalization of the relevant considerations to the realistic case of three generations is straightforward.

Thus, let us return to the GIM interaction Lagrangian (5.136) or (5.137). Now we are going to consider the "diagonal" processes of the type $q\bar{q} \to W^- W^+$, where q is a quark (u, d, s or c). In analogy with the results obtained in Section 5.3 for the electroweak interactions of leptons one may expect that in the quark sector one will also have to introduce (flavour-conserving) neutral currents and the corresponding interactions mediated by the neutral vector boson Z. First we will examine tree-level diagrams of the process $u\bar{u} \to W^+ W^-$. Contributions of the weak charged-current interaction (5.136) and of the electromagnetic interaction are depicted in Fig. 36.

Let us suppose that both final-state W's have longitudinal polarizations. In the same way as e.g. in the process $e^+ e^- \to W_L^+ W_L^-$, one has to add further diagrams to Fig. 36 if the tree unitarity is to be satisfied for the considered $u\bar{u}$ annihilation process. The diagrams necessary for a cancellation of quadratic and linear high-energy divergences arising in the contribution of Fig. 36 are shown in Fig. 37. The diagram in Fig. 37(a) contains a vertex corresponding to an interaction of quark neutral current with the Z. Taking into account the result (5.37) derived earlier for leptons, it is convenient to parametrize such an interaction for an arbitrary fermion f as follows:

$$\mathcal{L}_{ffZ} = \frac{g}{\cos\vartheta_W}\left(\varepsilon_L^{(f)}\bar{f}_L\gamma_\mu f_L + \varepsilon_R^{(f)}\bar{f}_R\gamma_\mu f_R\right)Z^\mu \tag{5.141}$$

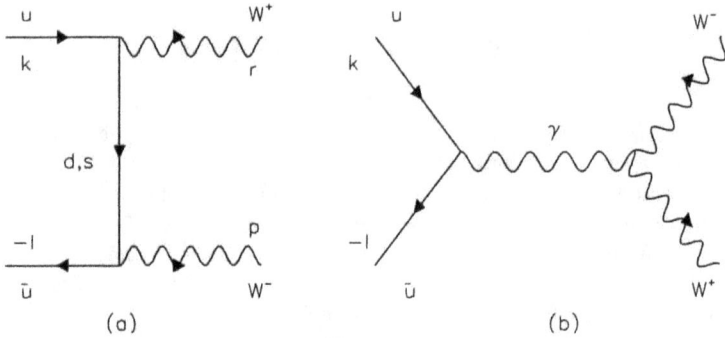

Fig. 36. *Tree-level diagrams of the process* $u\bar{u} \to W^+W^-$ *corresponding to weak charged current interactions (a) and the electromagnetic interaction (b).*

The constants $\varepsilon_{L,R}^{(f)}$, (which characterize separately the strength of the interaction of left-handed or right-handed fermions with the Z) may now be determined for the u-quark from the requirement of a cancellation of quadratic divergences arising in the limit $E \to \infty$ from the individual diagrams in Fig. 36, 37. Thus we obtain the following equations:

$$\frac{1}{2}g^2 \cos^2 \vartheta_C + \frac{1}{2}g^2 \sin^2 \vartheta_C - Q_u e^2 - \frac{g}{\cos \vartheta_W}\varepsilon_L^{(u)}g_{WWZ} = 0 \qquad (5.142)$$

$$- Q_u e^2 - \frac{g}{\cos \vartheta_W}\varepsilon_R^{(u)}g_{WWZ} = 0 \qquad (5.143)$$

(cf. (5.24), (5.25)). The relations (5.142) and (5.143) are written in a form which should make the origin of the individual terms obvious. We will only add several technical remarks: The last term on the left-hand side of Eq. (5.142) (which comes from Fig. 37(a)) has an opposite sign to the first two terms (which come from Fig. 36(a)) while in an analogous equation (5.24) the corresponding terms (i.e. the first and the last) have the same sign. Such a difference is due to the interchange of the external $W^\pm$ lines in Fig. 36(a) as compared to Fig. 17(a), which of course is related to the values of the relevant quark charges; in this sense, a natural counterpart of the process $u\bar{u} \to W^+W^-$ in the lepton sector is $\nu\bar{\nu} \to W^+W^-$ (cf. Fig. 16 and eq. (5.19)). In the electromagnetic contribution in (5.142), (5.143) we have made explicit the charge factor Q_u (for a comparison with (5.24) and (5.25) let us remember that $Q_e = -1$). If we now use the relations

$$g_{WWZ} = g \cos \vartheta_W$$
$$e = g \sin \vartheta_W$$

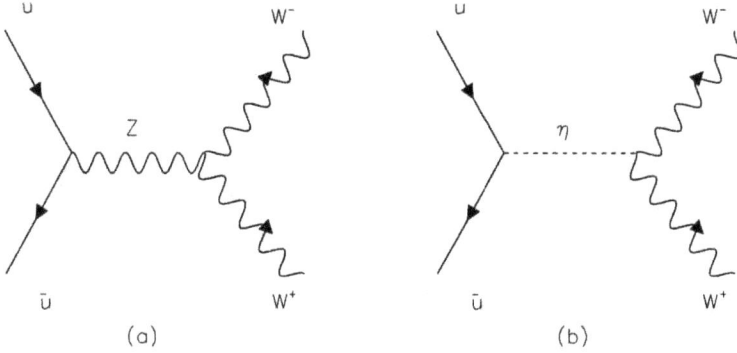

Fig. 37. The diagrams compensating the bad high-energy behaviour of the contribution of Fig. 36.

(see (5.36) and (5.37)), the solution of equations (5.142), (5.143) is obtained immediately:

$$\varepsilon_L^{(u)} = \frac{1}{2} - Q_u \sin^2 \vartheta_W \tag{5.144}$$

$$\varepsilon_R^{(u)} = -Q_u \sin^2 \vartheta_W \tag{5.145}$$

After the elimination of quadratically divergent asymptotic terms in the diagrams in Fig. 36 and 37(a) there still remains (similarly as in the process $e^+ e^- \to W_L^+ W_L^-$) a linear divergence:

$$\mathcal{M}^{(d)} + \mathcal{M}^{(s)} + \mathcal{M}^{(\gamma)} + \mathcal{M}^{(Z)} = -\frac{g^2}{4m_W^2} m_u \bar{v}(l) u(k) + O(1) \tag{5.146}$$

(the notation in the left-hand side of eq. (5.146) should be self-explanatory). The linearly divergent term in (5.146) is cancelled by the contribution of the diagram in Fig. 37(b). This graph contains a vertex corresponding to a Yukawa-type interaction, which for an arbitrary fermion f will be written as

$$\mathcal{L}_{ff\eta} = g_{ff\eta} \bar{f} f \eta \tag{5.147}$$

(cf. (5.75)). By means of manipulations analogous to those which in Section 5.5 have led from Eq. (5.74) to (5.77) one then finds that the required cancellation of linearly divergent terms occurs if and only if

$$g_{uu\eta} = -\frac{g}{2} \frac{m_u}{m_W} \tag{5.148}$$

One may proceed in a similar way for other processes of the considered type. An analysis of the tree diagrams for the process $d\bar{d} \rightarrow W_L^- W_L^+$ thus leads to the result (cf. (5.144), (5.145))

$$\varepsilon_L^{(d)} = -\frac{1}{2} - Q_d \sin^2 \vartheta_W \tag{5.149}$$

$$\varepsilon_R^{(d)} = -Q_d \sin^2 \vartheta_W \tag{5.150}$$

The different form of Eq. (5.149) as compared to Eq. (5.144) (i.e. the difference in the sign of the numerical constant $1/2$ in both expressions) is explained by the remark following the relation (5.143) (a "leptonic counterpart" of the process $d\bar{d} \rightarrow W^- W^+$ is $e^- e^+ \rightarrow W^- W^+$ — cf. the result for the g_L in (5.37)). From what we have already said it is also clear that for the s-quark neutral-current interaction one gets a result completely analogous to the d-quark case, i.e. the relations (5.149), (5.150), in which Q_d is replaced by Q_s (of course, $Q_d = Q_s = -1/3$ anyway). Similarly, for the c-quark we obtain the same formulae as in the u-quark case (i.e. (5.144), (5.145) with Q_u replaced by Q_c). These results can easily be generalized to the case of six quarks; for all the quarks with charge $-1/3$ (i.e. d, s, b) one obviously gets formulae of the type (5.149), (5.150), and for quarks with charge $2/3$ (i.e. u, c, t) formulae of the type (5.144), (5.145) are valid. Instead of the parameters $\varepsilon_{L,R}^{(f)}$ it is often convenient to employ their combinations v_f, a_f which characterize strengths of the interaction of the vector or axial-vector part of the neutral current with the Z. We will define the parameters v_f, a_f for an arbitrary fermion in an immediate analogy with (5.99); comparing it with (5.141) we define

$$v_f = \frac{1}{2}\left(\varepsilon_L^{(f)} + \varepsilon_R^{(f)}\right)$$

$$a_f = \frac{1}{2}\left(\varepsilon_L^{(f)} - \varepsilon_R^{(f)}\right) \tag{5.151}$$

Using the preceding results (see (5.144), (5.145), (5.149), (5.150))it is easy to find that

$$v_f = +\frac{1}{4} - Q_f \sin^2 \vartheta_W \qquad \text{for} \qquad f = u, c, t \tag{5.152}$$

$$v_f = -\frac{1}{4} - Q_f \sin^2 \vartheta_W \qquad \text{for} \qquad f = d, s, b \tag{5.153}$$

$$a_f = +\frac{1}{4} \qquad \text{for} \qquad f = u, c, t \tag{5.154}$$

$$a_f = -\frac{1}{4} \qquad \text{for} \qquad f = d, s, b \tag{5.155}$$

Let us recall (see (5.37) and (5.100)) that the formulae (5.153) and (5.154) are also valid for any neutrino, i.e. for $f = \nu_l$, $l = e, \mu, \tau$ (in this case of course $Q_f = 0$) and the formulae (5.153), (5.155) hold for an arbitrary charged lepton $l = e, \mu, \tau$ (in such a case $Q_l = -1$).

Finally, in all considered cases the coupling constant of the relevant Yukawa interaction (5.147) is given by

$$g_{ff\eta} = -\frac{g}{2}\frac{m_f}{m_W}.$$ (5.156)

(cf. (5.148)).

As regards other binary processes involving quarks, which are "potentially dangerous" from the point of view of the high-energy behaviour of the corresponding tree diagrams, such as e.g. $\bar{u}d \rightarrow W_L^- Z_L$, $\bar{u}d \rightarrow W_L^- \eta$, $u\bar{u} \rightarrow Z_L Z_L$ etc., it is not difficult to realize that the formulae (5.152) (5.156) together with (5.39) and the "universal" formulae for g_{WWZ}, $g_{WW\eta}$, $g_{ZZ\eta}$ (see (5.37), (5.73), (5.82)) already guarantee the tree unitarity to hold in the quark sector on the basis of mechanisms completely analogous to those discussed in detail for leptonic interactions.

Now we are in a position to discuss the last problem which remains to be solved, which however is of fundamental importance for the renormalizable theory of weak interactions: One has to find out what is the contribution of closed quark loops to the ABJ triangle anomaly which we have examined in Section 5.6 using a particular example within the framework of the leptonic sector of the theory. We have indicated that the quark and lepton contributions to the anomaly cancel each other; now we are going to prove this statement directly at least for the particular configuration of interaction vertices in the anomalous triangular fermion loop corresponding to the process discussed in Section 5.6.

Thus, let us consider the contribution to the ABJ anomaly coming from triangular closed loops in Fig. 30 when their internal lines correspond to an arbitrary fermion f. It is clear that the relation (5.119) giving a relevant numerical coefficient in the contribution of the electron loop to the anomaly may immediately be generalized; namely, for an arbitrary fermion f one may write

$$C_{anomaly}^{(f)} = a_f Q_f^2,$$ (5.157)

where Q_f is the corresponding charge factor and a_f is an axial-vector neutral-current interaction constant (see (5.151) and (5.154), (5.155)). We will now calculate the total contribution of quarks and leptons to the anomaly, according to (5.157), separately for each fermion generation (since the relevant properties of generations repeat themselves, it is easy to see that we always get the same result for different generations). With the help of (5.100), (5.154) and (5.155) we then get from (5.157) e.g. for the first generation of fermions (ν_e, e, u, d):

$$C_{anomaly}^{(lepton)} = -\frac{1}{4}$$ (5.158)

$$\begin{aligned} C_{anomaly}^{(quark)} &= \left(\frac{1}{4}Q_u^2 - \frac{1}{4}Q_d^2\right)N_c \\ &= \frac{1}{4}\cdot\left[\left(\frac{2}{3}\right)^2 - \left(-\frac{1}{3}\right)^2\right].3 = \frac{1}{4} \end{aligned}$$ (5.159)

where $N_c = 3$ is the number of quark colours. Let us recall that the term "colour" refers to an extension of the number of quark types - each flavour in fact corresponds to a triplet of quark fields distinguished by a "colour". Such a degree of freedom is irrelevant for the dynamics of electroweak interactions and that is why we have not considered it so far (it is however essential in the strong interaction dynamics - see e.g. [25]). Nevertheless, when adding contributions of the corresponding closed loops, one has to take into account all types of fermion fields including the quark colour. From (5.158) and (5.159) it is seen that

$$C^{(lepton)}_{anomaly} + C^{(quark)}_{anomaly} = 0, \qquad (5.160)$$

i.e. the contributions of quark and lepton loops to the ABJ anomaly cancel each other and such a cancellation occurs *separately for each generation*. Our earlier statement is thus proved. It is also interesting to notice that an essential point in the proof of eq. (5.160) is that the number of colours $N_c = 3$. This number is of course well substantiated experimentally in other situations (see e.g. [25]) and it is gratifying that results from different areas of particle physics sustain each other. We will also show that eq. (5.160) is equivalent to a remarkably simple identity for charges of the fermions belonging to the same generation. To this end, let us include formally in the lepton part the (vanishing) neutrino contribution as well; we thus get first

$$
\begin{aligned}
C^{(lepton)}_{anomaly} &+ C^{(quark)}_{anomaly} = \\
&= \frac{1}{4}Q_\nu^2 - \frac{1}{4}Q_e^2 + N_c\left(\frac{1}{4}Q_u^2 - \frac{1}{4}Q_d^2\right) \\
&= \frac{1}{4}\left[(Q_\nu - Q_e)(Q_\nu + Q_e) + N_c(Q_u - Q_d)(Q_u + Q_d)\right] \qquad (5.161)
\end{aligned}
$$

However,

$$Q_\nu - Q_e = Q_u - Q_d = 1 \qquad (5.162)$$

and from Eq. (5.161) and (5.162) it is obvious that Eq. (5.160) is equivalent to the identity

$$Q_\nu + Q_e + N_c(Q_u + Q_d) = 0 \qquad (5.163)$$

(which is obviously valid if $N_c = 3$), or

$$\sum_f Q_f = 0 \qquad (5.164)$$

where the sum in (5.164) means a summation over *all* fermions belonging to the same generation, i.e. including quark colours. It is important to realize that the anomaly cancellation condition, equivalent to (5.164), represents *the only* theoretical argument correlating properties of quarks and leptons, i.e. it implies a lepton-quark symmetry which is very natural from an aesthetic point of view (let us emphasize that the arguments for introducing the c-quark or the t-quark to implement the GIM mechanism or its generalization concern the quark sector only, and tell us nothing about a quark-lepton symmetry).

So far we have proved the absence of the ABJ anomaly in one particular case, namely for the configuration in which there are two photon lines and one Z-line attached to the corresponding vertices of relevant fermion loops. However, there are several other configurations where the ABJ triangle anomaly could play a role. If we denote the corresponding configuration by means of the triplet of vector bosons whose lines are attached to vertices of an anomalous triangular fermion loop, then we have – apart from the configuration $Z\gamma\gamma$ discussed earlier – the following additional possibilities: $ZZ\gamma$, ZZZ, ZWW, and γWW. Moreover, it is well known (see e.g. [17], [21], [25], [48]), that the ABJ anomalies occur in triangular fermion loops of two types:

VVA (two vector vertices and one of the axial-vector type),

AAA (three axial-vector vertices).

Of course, in the configuration $Z\gamma\gamma$ considered up to now only the VVA fermion loops play a role (and the same is true in the $ZZ\gamma$ case) but in configurations ZZZ and WWZ one has to consider both the VVA and the AAA fermion loops.

One can demonstrate that the ABJ triangle anomalies vanish (i.e. cancel) *in all* the above-mentioned cases; again, the mechanism described in the case of the $Z\gamma\gamma$ configuration plays an essential role. In other words, the contributions coming from quark and lepton anomalous triangle loops cancel each other owing to the identity (5.164) (absence of some anomalies is however trivial). A proof of the complete cancellation of anomalies within the standard electroweak theory is left to the interested reader as an instructive exercise (see the problem 5.15).

It is remarkable that a complete cancellation of the ABJ anomalies occurs automatically, as a consequence of properties of the electroweak interactions of quarks and leptons (which have been deduced from the requirement of the tree unitarity) and of a choice of the quark charge spectrum which is very natural from a physical point of view. Anyway, the elimination of anomalies is technically the last crucial step in the construction of an internally consistent model of electroweak interactions.

Now we have come to an end of our road to the renormalizable theory of weak and electromagnetic interactions. The last "missing link" in the electroweak Lagrangian (5.95) is the interaction of charged and neutral quark currents with the vector bosons $W^{\pm}$ and Z described by the expressions (5.137) (or (5.139)), (5.141), (5.151) — (5.155) and the Yukawa interaction of quarks with the scalar field η (see (5.147),(5.156)) and of course a standard electromagnetic interaction of quarks. The final result of our construction corresponds precisely to the Lagrangian of the standard model of electroweak interactions which is currently (together with quantum chromodynamics) one of the cornerstones of modern particle theory (see [25]). The full interaction Lagrangian which we have obtained is for convenience summarized in Appendix K. The method of deriving the standard model of electroweak interactions, which we have described in this chapter, relied substantially on the criterion of tree unitarity; as we have seen, the elimination of anomalies is then an automatic consequence of a physically realistic choice of the quark sector of the model. The absence of manifest sources of undesirable divergences in perturbation expansion indicates that the model we have obtained is renormalizable (cf. the discussion at the end of Section 5.5). It turns out that such a guess is indeed correct: Now we have an interaction Lagrangian which leads to a renormalizable perturbation expansion for the S-matrix. However, a proof of such a statement is far from straightforward; for carrying out the corresponding proof to all orders of perturbation expansion it was necessary to reformulate non-trivially the whole theory and to apply some remarkable new techniques and methods of quantum gauge field theory (see [10]). A technical discussion of these problems can be found in many textbooks and review articles (see e.g. [15], [17], [21], [25]).

The derivation of the standard model of electroweak interactions described in this chapter is remarkable in particular because it demonstrates the *necessity* of introducing vector bosons and interactions of the Yang-Mills type (this corresponds to the principle of non-abelian gauge invariance in the traditional GWS formulation) and at least one elementary scalar boson (which corresponds to the GWS realization of the Higgs mechanism) if one wants to arrive at a renormalizable theory of weak and electromagnetic interactions. In other words, and in a more detailed way: After the formulation of the GWS theory [5 - 7] one might naturally contemplate the question of whether one could do without the Higgs scalar boson (whose presence is somewhat "uncomfortable" - see below). The systematic deductive approach [11 14] described in some detail in the preceding section shows that the ingenious GWS construction based on principles of broken symmetry in fact represents *the only realistic possibility* for a renormalizable electroweak unification, if at the same time we restrict the number of possible new particles (i.e. if we have in mind a "minimal" model); let us recall again that in comparison with the naive "electro-weak" theory (4.26) we had to introduce (within such a minimal strategy) one extra neutral vector boson and one neutral scalar boson.

As we have already mentioned, the assumed existence of a neutral scalar Higgs boson is somewhat uncomfortable; by that we mean, in particular, that the standard model does not predict any specific value of a mass for such a particle (in contrast to the case of an IVB). On the other hand, we have also mentioned that the requirement of *perturbative* renormalizability is a restriction of rather technical nature from a modern point of view (see e.g. [72]), and its physical relevance is not, strictly speaking, quite clear. The problem of the "Higgs sector" of electroweak interactions thus represents one of the most interesting open questions of contemporary particle physics (see e.g. [73], [74]). One may expect that this intriguing problem will be elucidated by the planned experiments on LEP 200 and LHC (CERN) which, moreover, should also show whether the self-interactions of vector bosons are indeed of the Yang-Mills type. It is supposed that the corresponding tests of these fundamental aspects of the standard model will be feasible in the foreseeable future — by the end of this (or at the beginning of the next) millenium.

Problems

5.1. Derive (5.10).

5.2. Prove the statement following the relation (5.19).

5.3. Derive (5.50).

5.4. Derive in detail (5.53) (a sketch of the proof is given in Appendix J).

5.5. Derive (5.56).

5.6. Derive (5.59), (5.61) and (5.62).

5.7. Derive (5.72).

5.8. Derive (5.76).

5.9. Derive (5.78).

5.10. Derive (5.80).

5.11. Derive (5.83) and (5.85).

5.12. Prove that scattering amplitudes of the processes (5.97) corresponding to the interaction Lagrangian (5.95) satisfy the condition of tree unitarity.

5.13. Derive (5.123) and (5.127).

5.14. Show that when there exist five quarks u, d, s, c, b, a sixth quark t with charge 2/3 is also necessary if one wants to suppress bad high-energy behaviour of all relevant tree-level scattering amplitudes and, at the same time, to avoid flavour-changing neutral currents. Prove that the matrix V_{CKM} on (5.139), (5.140) must then be unitary. How can one arrive at a parametrization mentioned in the text following (5.140)?

5.15. Prove that the ABJ triangle anomalies vanish (for an arbitrary fermion generation) also in configurations $ZZ\gamma$, ZZZ, ZWW and γWW (in the sense defined at the end of Section 5.7). In doing this, neglect the mixing of different generations in quark sector. How does a non-trivial mixing influence the cancellation of anomalies?

5.16. Calculate (in tree approximation) cross sections of elastic scattering processes $\nu_\mu e \to \nu_\mu e$ and $\bar\nu_\mu e \to \bar\nu_\mu e$. Discuss separately the low-energy and high-energy regions. Explain how could one determine, from a measurement of cross sections $\sigma(\nu_\mu e \to \nu_\mu e)$ and $\sigma(\bar\nu_\mu e \to \bar\nu_\mu e)$ in *low-energy* region (i.e. for $s \ll G_F^{-1}$) the neutral-current parameter $sin^2\vartheta_W$ and how could one verify validity of the Weinberg relation $m_W/m_Z = \cos\vartheta_W$ (*predicting* thus the values m_W and m_Z without a direct detection of W and Z).

5.17. Calculate (in tree approximation) cross sections of processes $\nu_e e \to \nu_e e$ and $\bar\nu_e e \to \bar\nu_e e$.

     a) In the low-energy region compare the obtained results with Feynman – Gell-Mann theory (see Appendix D).

     b) Does $\sigma(\nu_e e \to \nu_e e) = \sigma(\bar\nu_e e \to \bar\nu_e e)$ in the limit $s \to \infty$?

5.18. Calculate (in tree approximation)

     a) total decay width of the W

     b) total decay width of the Z

     c) the decay width $\Gamma(\eta \to f\bar f)$ where η is the Higgs scalar boson and f is an arbitrary fermion (such that $2m_f < m_\eta$). What is the ratio of lepton and hadron (i.e. quark) widths in the case of W, Z and η decays assuming that $m_\eta = 300$ GeV and $m_t = 120$ GeV?

     d) For the value of m_η considered above calculate also the decay widths $\Gamma(\eta \to W^-W^+)$ and $\Gamma(\eta \to ZZ)$. For what value of m_η is the decay width of the scalar boson η comparable with its mass?

5.19. Let us imagine that the electromagnetic interaction is switched off, i.e. $e = 0$ (in such a hypothetical world an electron differs from the corresponding neutrino only by its rest mass). Is it possible then to construct a renormalizable theory of weak interactions incorporating the original naive model with W bosons? Does an anomaly cancellation condition lead to a restriction on the fermion spectrum? How can one interpret a role of unification of weak and electromagnetic interactions in constructing a corresponding renormalizable theory in the realistic case $e \neq 0$?

Appendix A

KINEMATICS

In this appendix we have summarized some formulae of relativistic kinematics which are needed in the main text.

For a binary reaction $1 + 2 \to 3 + 4$ let us denote the four-momenta of particles $1,\ldots 4$ (with rest masses $m_1, \ldots, m_4$) consecutively as k, p, k', p', so that

$$k + p = k' + p' \tag{A.1}$$

and

$$k^2 = m_1^2, \quad p^2 = m_2^2, \quad k'^2 = m_3^2, \quad p'^2 = m_4^2 \tag{A.2}$$

If one defines the standard Mandelstam variables (kinematical invariants) as

$$
\begin{aligned}
s &= (k + p)^2 \\
t &= (k - k')^2 \\
u &= (k - p')^2
\end{aligned}
\tag{A.3}
$$

then the following familiar relation holds:

$$s + t + u = \sum_{j=1}^{4} m_j^2 \tag{A.4}$$

The identity (A.4) is most easily proved as follows:

According to the definition (A.3) and using the four-momentum conservation (A.1) one may write

$$
\begin{aligned}
s + t + u &= \frac{1}{2}[(k + p)^2 + (k' + p')^2 + (k - k')^2 + (p - p')^2 + (k - p')^2 + (k' - p)^2] \\
&= k^2 + p^2 + k'^2 + p'^2 + \frac{1}{2}(k + p - k' - p')^2
\end{aligned}
$$

From the last expression and from (A.1), (A.2) then the result (A.4) follows immediately. Let us recall that $s = E_{c.m.}^2$, where $E_{c.m.}$ is the total energy of colliding particles in the centre-of-mass (c.m.) system.

Further, we will introduce a dimensionless variable

$$y = \frac{p.q}{p.k} \tag{A.5}$$

where we have denoted $q = k - k'$. This kinematical variable is particularly useful in cross-section calculations in situations when one may neglect rest masses of particles. In a massless case the following relations hold (we leave a corresponding proof to the reader as a simple exercise)

$$
\begin{aligned}
t &= -sy \\
u &= -s(1 - y)
\end{aligned} \tag{A.6}
$$

Moreover, the variable y in such a case is simply related to the scattering angle in the c.m. system:

$$y = \frac{1}{2}(1 - \cos \vartheta) \tag{A.7}$$

where ϑ is defined as the angle between momenta $\vec{k}$ and $\vec{k}'$. A proof of (A.7) is easy if one takes into account that upon neglecting masses one has in the c.m. system

$$k_o = |\vec{k}| = p_o = |\vec{p}| = k'_o = |\vec{k}'| = p'_o = |\vec{p}'| = E = \frac{1}{2}\sqrt{s}$$

Then

$$
\begin{aligned}
p.q &= p.k - p.k' \\
&= p.k - [E^2 - E^2 \cos(\pi - \vartheta)] \\
&= p.k - \frac{1}{4}s(1 + \cos \vartheta)
\end{aligned}
$$

However, one also has $p.k = \frac{1}{2}s$ and from the definition (A.5) we thus immediately obtain (A.7). It is also obvious from (A.7) that if one neglects masses, the variable y takes on values from 0 to 1.

We will now give two frequently used formulae. The first expresses the momentum of particles colliding in the c.m. system, as a function of the kinematical invariant s and of the relevant rest masses m_1, m_2:

$$|\vec{p}_{c.m.}| = \left[\frac{\lambda(s, m_1^2, m_2^2)}{4s}\right]^{\frac{1}{2}} \tag{A.8}$$

where

$$\lambda(x, y, z) = x^2 + y^2 + z^2 - 2xy - 2xz - 2yz \tag{A.9}$$

The proof of (A.8) is straightforward. Total energy of the two particles in the c.m. system (using the shorthand notation $\vec{p}_{c.m.} = \vec{p}$) is

$$\sqrt{\vec{p}^2 + m_1^2} + \sqrt{\vec{p}^2 + m_2^2} = \sqrt{s} \tag{A.10}$$

Solving eq. (A.10) with respect to $|\vec{p}|$ we get first

$$|\vec{p}|^2 + m_1^2 = \left(\sqrt{s} - \sqrt{|\vec{p}|^2 + m_2^2} \right)^2$$

from where (A.8) follows after a short manipulation.

The second frequently used formula gives the magnitude of relative velocity of 2 particles in a collision (in an arbitrary reference frame). Let the two colliding particles with rest masses m_1, m_2 have antiparallel velocities $\vec{v}_1$, $\vec{v}_2$. Then

$$|\vec{v}_1 - \vec{v}_2| = \frac{[(p_1 \cdot p_2)^2 - m_1^2 m_2^2]^{\frac{1}{2}}}{E_1 E_2} \tag{A.11}$$

where $p_i = (E_i, \vec{p}_i)$, $i = 1, 2$ are four-momenta of particles *1,2*, i.e. $p_i^2 = m_i^2$. The proof of (A.11) is easy: Under given conditions one has

$$\begin{aligned} |\vec{v}_1 - \vec{v}_2| &\equiv \sqrt{(\vec{v}_1 - \vec{v}_2)^2} = \sqrt{\left(\frac{|\vec{p}_1|}{E_1} + \frac{|\vec{p}_2|}{E_2} \right)^2} \\ &= \frac{1}{E_1 E_2} \sqrt{(E_1 E_2 + |\vec{p}_1| \cdot |\vec{p}_2|)^2 - m_1^2 m_2^2} \end{aligned}$$

(the last identity becomes clear if we use $|\vec{p}_i| = \sqrt{E_i^2 - m_i^2}$). However, in the considered configuration of the particle momenta one may write

$$p_1 \cdot p_2 = E_1 E_2 + |\vec{p}_1||\vec{p}_2|$$

and the relation (A.11) is thus proved. The formula (A.11) may be also recast in terms of the function $\lambda(s, m_1^2, m_2^2)$ introduced in (A.9). Indeed, from the definition (A.3) it follows that

$$p_1 \cdot p_2 = \frac{1}{2}(s - m_1^2 - m_2^2)$$

and substituting this to (A.11) we get immediately

$$|\vec{v}_1 - \vec{v}_2| = \frac{\lambda^{\frac{1}{2}}(s, m_1^2, m_2^2)}{2 E_1 E_2} \tag{A.12}$$

Finally, using (A.8) one may also write

$$|\vec{v}_1 - \vec{v}_2| = \frac{s^{\frac{1}{2}} |\vec{p}_{c.m.}|}{E_1 E_2} \tag{A.13}$$

Appendix B

DIRAC SPINORS AND FREE FIELDS

External lines of Feynman diagrams corresponding to spin-$\frac{1}{2}$ fermions represent graphically solutions of the Dirac equation in momentum space (for a four-momentum p we always take $p_0 = +\sqrt{\vec{p}^2 + m^2}$):

$$(\not{p} - m)u = 0, \quad (\not{p} + m)v = 0 \tag{B.1}$$

The u, v in (B.1) is a shorthand notation for $u(p,s)$, $v(p,s)$, where s is a polarization which takes on 2 possible values. The symbol $\not{p}$ in (B.1) is defined as $\not{p} = p_\mu \gamma^\mu$ where $\gamma^\mu, \mu = 0, 1, 2, 3$ are standard Dirac matrices.

In diagrams, a factor of u (or $\bar{u}$) corresponds to a particle, and similarly v (or $\bar{v}$) stands for an antiparticle. From (B.1) it follows immediately that for conjugated spinors $\bar{u}, \bar{v}$ one has (recall that $\bar{u} = u^\dagger \gamma_0$, $\bar{v} = v^\dagger \gamma_0$)

$$\bar{u}(\not{p} - m) = 0, \quad \bar{v}(\not{p} + m) = 0 \tag{B.2}$$

The functions u, v are normalized by

$$\bar{u}u = 2m, \quad \bar{v}v = -2m \tag{B.3}$$

If we use the convention (B.3), an expansion of a free Dirac field in plane waves may be written as

$$
\begin{aligned}
\psi(x) &= \sum_{\pm s} \int \frac{d^3 p}{(2\pi)^{\frac{3}{2}}(2p_0)^{\frac{1}{2}}} [b(p,s)u(p,s)e^{-ipx} + d^+(p,s)v(p,s)e^{ipx}] \\
\bar{\psi}(x) &= \sum_{\pm s} \int \frac{d^3 p}{(2\pi)^{\frac{3}{2}}(2p_0)^{\frac{1}{2}}} [b^+(p,s)\bar{u}(p,s)e^{ipx} + d(p,s)\bar{v}(p,s)e^{-ipx}]
\end{aligned}
\tag{B.4}
$$

where b (b^+) is an annihilation (creation) operator for a particle and d (d^+) correspond to antiparticles. Let us remark that the annihilation and creation operators in (B.4) satisfy anticommutation relations

$$\{b(p,s), b^+(p's')\} = \{d(p,s), d^+(p',s')\} = \delta_{ss'}\delta^3(\vec{p} - \vec{p'})$$

etc. which correspond to the normalization of one-particle states defined by

$$< \vec{p}, s | \vec{p}', s' > = \delta_{ss'} \delta^3(\vec{p} - \vec{p}')$$

It is in order to emphasize here that, instead of the convention (B.3), another normalization is frequently used in the literature, namely $\bar{u}u = 1$, $\bar{v}v = -1$ (see e.g. [16], [21]). An advantage of the option (B.3) is that the relevant formula for a scattering cross section or a decay rate then has the same form both for bosons and fermions (see Appendix C, formulae (C.1) or (C.14) resp.) and that a Lorentz-invariant scattering amplitude $\mathcal{M}_{fi}$ for an arbitrary *binary* process is *dimensionless* (cf. (C.3)).

For the functions u, v normalized according to (B.3) one has further

$$\sum_{\pm s} u(p, s)\bar{u}(p, s) = \not{p} + m \tag{B.5}$$

$$\sum_{\pm s} v(p, s)\bar{v}(p, s) = \not{p} - m \tag{B.6}$$

Finally, let us specify an explicit form of the functions u, v satisfying (B.1), (B.3). We will denote $u(p, s) \equiv u^{(r)}(p)$, i.e. the polarizations $\pm s$ are labelled by an index $r = 1, 2$. Solutions of (B.1) with polarizations corresponding to definite projections of the spin onto the z-axis in the *rest frame* of the considered particle (we assume $m \neq 0$) are given by

$$u^{(r)}(p) = \sqrt{E + m} \begin{pmatrix} \chi^{(r)} \\ \dfrac{\vec{\sigma} \cdot \vec{p}}{E + m} \chi^{(r)} \end{pmatrix} \tag{B.7}$$

$$v^{(r)}(p) = \pm\sqrt{E + m} \begin{pmatrix} \dfrac{\vec{\sigma} \cdot \vec{p}}{E + m} \chi^{(r)} \\ \chi^{(r)} \end{pmatrix} \tag{B.8}$$

In (B.7), (B.8) we have denoted

$$\chi^{(1)} = \begin{pmatrix} 1 \\ 0 \end{pmatrix}, \quad \chi^{(2)} = \begin{pmatrix} 0 \\ 1 \end{pmatrix}$$

and $\vec{\sigma}$ are Pauli matrices. The upper sign in (B.8) refers to $r = 1$, the lower sign corresponds to $r = 2$. The signs $\pm$ in (B.8) are chosen so that the operation of charge conjugation would turn a function u into a v, assuming that a phase of the charge-conjugation matrix is fixed conventionally ($C = i\gamma^2\gamma^0$).

It is important to notice that in the ultrarelativistic limit (i.e. for $E \gg m$) the u and v behave like $\sqrt{E}$; this fact is frequently used in estimates of high-energy asymptotics of scattering amplitudes represented in terms of Feynman diagrams.

Appendix C

FORMULAE FOR CROSS SECTIONS AND DECAY RATES

If we fix our conventions so that the Dirac spinors $u, \bar{u}, v, \bar{v}$ corresponding to external fermion lines in Feynman diagrams are normalized according to the (B.3), then a general formula for the differential cross section of a process $1 + 2 \rightarrow 3 + \cdots + n$ (cf. [16]) reads

$$d\sigma = \frac{1}{|\vec{v}_1 - \vec{v}_2|} \frac{1}{2E_1} \frac{1}{2E_2} |\mathcal{M}_{fi}|^2 (2\pi)^4 \delta^4(p_1 + p_2 - \sum_{j=3}^{n} p_j) \frac{d^3 p_3}{(2\pi)^3 2E_3} \cdots \frac{d^3 p_n}{(2\pi)^3 2E_n} K \quad \text{(C.1)}$$

regardless of whether the particles $1,2,\ldots\,n$ are bosons or fermions. In the formula (C.1) we have denoted by $\vec{v}_1, \vec{v}_2$ velocities of the initial particles $1,2$ (we take them to be parallel and of opposite directions), the $p_j, j = 1, \ldots, n$ are on-shell four-momenta and E_j denote the corresponding energies, i.e. $E_j = \sqrt{\vec{p}_j^2 + m_j^2}$ for $j = 1, \ldots, n$ and K is a combinatorial (statistical) factor, which is different from 1 only when some of the final-state particles $3,\ldots,n$ are identical, namely

$$K = \prod_{r=1}^{k} \frac{1}{n_r!} \quad \text{(C.2)}$$

where n_r is the number of identical particles of the r-th kind in the final state (of course, $n_1 + \ldots + n_k = n - 2$). The $\mathcal{M}_{fi}$ is a relativistic invariant scattering amplitude which in practice is calculated as the contribution of Feynman diagrams relevant for the considered process. Let us remark that owing to the employed normalization of one-particle states (corresponding to (B.4)) the $\mathcal{M}_{fi}$ is connected with the corresponding S-matrix element via the relation

$$S_{fi} = \delta_{fi} + (2\pi)^4 \delta^4(P_f - P_i)(i\mathcal{M}_{fi}) \prod_{f,i} \frac{1}{(2\pi)^{3/2}(2E_{f,i})^{1/2}}$$

where P_f or P_i is the total four-momentum of the final or initial particles. The convention used by Bjorken and Drell [16] differs from our definition by replacing $i\mathcal{M}_{fi} \to -i\mathcal{M}_{fi}$.

Using the formula (C.1) one may easily determine a dimension of the amplitude $\mathcal{M}_{fi}$. The dimension of the left-hand side of (C.1) is

$$[d\sigma] = M^{-2}$$

where M is an arbitrary mass and on the right-hand side of (C.1) one has

$$M^{-1}.M^{-1}[|\mathcal{M}_{fi}|^2].M^{-4}.(M^2)^{n-2} = [|\mathcal{M}_{fi}|^2].M^{2n-10}$$

(recall that the dimension of the four-dimensional delta function is M^{-4}!). Thus, for the dimension of $\mathcal{M}_{fi}$ we obtain the equation

$$M^{-2} = [|\mathcal{M}_{fi}|^2].M^{2n-10}$$

from where we get immediately

$$[\mathcal{M}_{fi}] = M^{4-n} \tag{C.3}$$

In particular, (C.3) implies that a scattering amplitude of an arbitrary *binary* process $1 + 2 \to 3 + 4$ (i.e. for $n = 4$) is *dimensionless* (let us stress again that the normalization convention (B.3) is crucial for such a statement to be valid). This simple fact is frequently used in the main text for estimates of high-energy behaviour of scattering amplitudes of weak and electromagnetic processes.

For the relative velocity $|\vec{v}_1 - \vec{v}_2|$ in (C.1) we may use formulae (A.11) or (A.12) from Appendix A and thus obtain commonly used equivalent alternatives to (C.1) in which the factor

$$|\vec{v}_1 - \vec{v}_2|^{-1}(2E_1)^{-1}(2E_2)^{-1}$$

is replaced by

$$\frac{1}{4}[(p_1.p_2)^2 - m_1^2 m_2^2]^{-\frac{1}{2}}$$

or by

$$\frac{1}{2}\lambda^{-\frac{1}{2}}(s, m_1^2, m_2^2)$$

Further, we are going to derive a practically useful formula for the differential cross section of a binary process with respect to the scattering angle in the centre-of-mass system of colliding particles. Let us consider a process $1 + 2 \to 3 + 4$ in the centre-of-mass (c.m.) system, i.e. take $\vec{p}_1 = -\vec{p}_2 = \vec{p}_{c.m.}$ and $E_1 + E_2 = \sqrt{s}$ (the $|\vec{p}_{c.m.}|$ is of course given by the formula (A.8) — see Appendix A). We will assume that the particles *3, 4* are not identical; in the opposite case we would just have to include a combinatorial factor $K = \frac{1}{2}$. From the general formula (C.1) we then get first (see also (A.13))

$$d\sigma = \frac{1}{4|\vec{p}_{c.m.}|s^{\frac{1}{2}}}|\mathcal{M}_{fi}|^2(2\pi)^4\delta^4(p_1 + p_2 - p_3 - p_4)\frac{d^3p_3}{(2\pi)^3 2E_3}\frac{d^3p_4}{(2\pi)^3 2E_4} \quad \text{(C.4)}$$

The relation (C.4) may now be integrated to eliminate the δ-function; in doing this, we will still use the same symbol $d\sigma$ for the integrated cross section. First of all, one may integrate trivially over d^3p_4 to get

$$d\sigma = \frac{1}{64\pi^2}\frac{1}{|\vec{p}_{c.m.}|s^{\frac{1}{2}}}|\mathcal{M}_{fi}|^2\delta\left(\sqrt{|\vec{p}|^2 + m_3^2} + \sqrt{|\vec{p}|^2 + m_4^2} - \sqrt{s}\right) \times$$
$$\times \frac{d^3p'}{\sqrt{|\vec{p}|^2 + m_3^2}\sqrt{|\vec{p}|^2 + m_4^2}} \quad \text{(C.5)}$$

where $\vec{p}' = \vec{p}_3 = -\vec{p}_4$; in (C.5) we have also set $E_1 + E_2 = \sqrt{s}$. A direction of the $\vec{p}$ may be described by spherical angles ϑ, φ (the axis 3 of the coordinate frame is defined by the $\vec{p}$ direction) and one may then write

$$d^3\vec{p}' = |\vec{p}|^2 d|\vec{p}'|d\Omega = |\vec{p}'|^2 d|\vec{p}'|\sin\vartheta d\vartheta d\varphi$$

Let us now integrate (C.5) with respect to $|\vec{p}'|$ (in the limits 0 and ∞); thus we get rid of the δ-function corresponding to energy conservation. For brevity, let us denote $|\vec{p}'| = z$. Using such a notation, (C.5) reads

$$d\sigma = \frac{1}{64\pi^2}\frac{1}{|\vec{p}_{c.m.}|s^{\frac{1}{2}}}|\mathcal{M}_{fi}|^2\delta[f(z)]\frac{z^2 dz d\Omega}{\sqrt{z^2 + m_3^2}\sqrt{z^2 + m_4^2}} \quad \text{(C.6)}$$

where

$$f(z) = \sqrt{z^2 + m_3^2} + \sqrt{z^2 + m_4^2} - \sqrt{s} \quad \text{(C.7)}$$

The equation $f(z_0) = 0$ has a single positive solution z_0, namely (see (A.8))

$$z_0 = |\vec{p}_{c.m.}| = \left[\frac{\lambda(s, m_3^2, m_4^2)}{4s}\right]^{\frac{1}{2}} \quad \text{(C.8)}$$

The δ-function in (C.6) is then equivalent to

$$\delta[f(z)] = \frac{1}{|f'(z_0)|}\delta(z - z_0) \quad \text{(C.9)}$$

From (C.7) it follows easily that

$$f'(z_0) = \frac{z_0}{\sqrt{z_0^2 + m_3^2}} + \frac{z_0}{\sqrt{z_0^2 + m_4^2}} = \frac{z_0\sqrt{s}}{\sqrt{z_0^2 + m_3^2}\sqrt{z_0^2 + m_4^2}} \quad \text{(C.10)}$$

Substituting (C.9) and (C.10) into (C.6), an integration of (C.6) with respect to z is trivial and by using (C.8) we obtain finally

$$\frac{d\sigma}{d\Omega} = \frac{1}{64\pi^2} \frac{1}{s} \frac{|\vec{p}_{c.m.}|}{|\vec{p}_{c.m.}|} |\mathcal{M}_{fi}|^2 \qquad (C.11)$$

Obviously, the formula (C.11) may also be recast as

$$\frac{d\sigma}{d\Omega} = \frac{1}{64\pi^2} \frac{1}{s} \frac{\lambda^{\frac{1}{2}}(s, m_3^2, m_4^2)}{\lambda^{\frac{1}{2}}(s, m_1^2, m_2^2)} |\mathcal{M}_{fi}|^2 \qquad (C.12)$$

When one may neglect particle masses it is useful to work with differential cross section (of a binary process) defined with respect to the Lorentz invariant dimensionless variable y defined in Appendix A (see(A.5)). If the $|\mathcal{M}_{fi}|^2$ depends only on the angle ϑ then using (A.7) one gets from (C.11) or (C.12) a simple formula

$$\frac{d\sigma}{dy} = \frac{1}{16\pi} \frac{1}{s} |\mathcal{M}_{fi}|^2 \qquad (C.13)$$

In practical calculations, the Mandelstam invariants t, u in $|\mathcal{M}_{fi}|^2$ may then be expressed in terms of s and y (see (A.6)). The integral cross section is then obtained by integrating (C.13) over the y from 0 to 1.

Let us now consider a two-particle decay of a particle with mass M in its rest frame; masses of the decay products will be denoted as m_1, m_2. The differential decay probability per unit time is given (cf. [16]) by

$$dw = \frac{1}{2M} |\mathcal{M}_{fi}|^2 (2\pi)^4 \delta^4(P - p_1 - p_2) \frac{d^3p_1}{(2\pi)^3 2E_1} \frac{d^3p_2}{(2\pi)^3 2E_2} K \qquad (C.14)$$

where $\mathcal{M}_{fi}$ is the corresponding relativistic invariant decay amplitude (determined by the relevant Feynman diagrams), P is the four-momentum of the decaying particle, i.e. (in the rest system) $P = (M, 0, 0, 0)$, $p_i = (E_i, \vec{p}_i)$ for $i = 1, 2$ are four-momenta of the final-state particles *1,2* and K is the combinatorial factor defined in (C.2). In what follows we will consider for simplicity the case $1 \neq 2$, i.e. $K = 1$.

The phase-space integration of the differential decay rate (C.14) (i.e. an integration over the momenta of the final-state particles *1,2* may be performed in analogy with the previous derivation of the formula (C.11). If we denote the integrated element of the two-particle phase-space volume corresponding to a solid-angle element $d\Omega$ by a symbol $d(LIPS_2)$ (where "*LIPS*" is an acronym for "Lorentz Invariant Phase Space") we thus obtain

$$\begin{aligned}
d(LIPS_2) &= \frac{1}{4\pi^2} \int \frac{d^3p_1}{2E_1} \frac{d^3p_2}{2E_2} \delta^4(P - p_1 - p_2) \\
&= \frac{d\Omega}{16\pi^2} \int_0^\infty dz \frac{z^2}{\sqrt{z^2 + m_1^2}\sqrt{z^2 + m_2^2}} \times \\
&\times \delta\left(\sqrt{z^2 + m_1^2} + \sqrt{z^2 + m_2^2} - M\right) = \frac{|\vec{p}|}{M} \frac{d\Omega}{16\pi^2}
\end{aligned} \qquad (C.15)$$

where $|\vec{p}|$ is the magnitude of three-momentum of a decay product (remember that $|\vec{p}_1| = |\vec{p}_2| = |\vec{p}|$). The $|\vec{p}|$ is of course given by (cf. (C.7), (C.8))

$$|\vec{p}| = \frac{1}{2M} \lambda^{\frac{1}{2}}(M^2, m_1^2, m_2^2) \tag{C.16}$$

Note that using the definition (A.9), the expression $\lambda(M^2, m_1^2, m_2^2)$ may be rewritten as

$$\lambda(M^2, m_1^2, m_2^2) = [M^2 - (m_1 + m_2)^2][M^2 - (m_1 - m_2)^2] \tag{C.17}$$

Thus, when it makes sense to consider an angular distribution of the decay products (e.g. if the decaying particle is polarized) we have a general formula for the corresponding differential decay rate

$$dw = \frac{1}{2M} |\mathcal{M}_{fi}|^2 d(LIPS_2) \tag{C.18}$$

where the element of the phase space is given by (C.15). If the initial and final-state particles are unpolarized, the quantity $|\mathcal{M}_{fi}|^2$ summed over polarizations does not depend on the angles $\Omega \equiv (\vartheta, \varphi)$ and the relation (C.18) may be integrated trivially; thus we get a useful formula for the integral decay rate (decay width) Γ:

$$\Gamma = \frac{1}{2M} \overline{|\mathcal{M}_{fi}|^2} LIPS_2 \tag{C.19}$$

where the symbol $\overline{|\mathcal{M}_{fi}|^2}$ indicates, as usual, summing and averaging over polarizations and the phase-space factor is

$$
\begin{aligned}
LIPS_2 &= \frac{1}{4\pi} \frac{|\vec{p}|}{M} \\
&= \frac{1}{8\pi} \sqrt{1 - \frac{(m_1 + m_2)^2}{M^2}} \sqrt{1 - \frac{(m_1 - m_2)^2}{M^2}}
\end{aligned}
\tag{C.20}
$$

The last expression follows easily from the relations (C.15) through (C.17).

For completeness we give finally two frequently used particular cases of the formula (C.20):

i) If $m_1 = m_2 = m$ we get from (C.20)

$$LIPS_2|_{m_1 = m_2 = m} = \frac{1}{8\pi} \sqrt{1 - \frac{4m^2}{M^2}} \tag{C.21}$$

ii) For $m_1, m_2 \ll M$ we have a very simple approximate formula

$$LIPS_2|_{m_1, m_2 \ll M} \doteq \frac{1}{8\pi} \tag{C.22}$$

Appendix D

NEUTRINO-ELECTRON SCATTERING

As an illustration of the considerations presented in Chapter 2, in this appendix we will perform a detailed calculation of cross sections of the elastic scattering processes $\nu_e e \to \nu_e e$ and $\bar{\nu}_e e \to \bar{\nu}_e e$ in the lowest perturbative order within a Fermi–type model of weak interactions. More precisely, we will employ the model of direct four-fermion interaction of the type current $\times$ current, with currents $V - A$ (see (2.1)), i.e. the classic Feynman – Gell-Mann theory [2]. The relevant Feynman diagrams are shown in Fig. 38. The Lorentz-invariant scattering amplitudes $\mathcal{M}_{fi}$ corresponding to the diagrams $(a), (b)$ in Fig. 38 are given by

$$i\mathcal{M}_{fi}^{(a)} = -i\frac{G_F}{\sqrt{2}}[\bar{u}(p')\gamma^\rho(1 - \gamma_5)u(k)]\,[\bar{u}(k')\gamma_\rho(1 - \gamma_5)u(p)] \qquad (D.1)$$

$$i\mathcal{M}_{fi}^{(b)} = -i\frac{G_F}{\sqrt{2}}[\bar{v}(k)\gamma^\rho(1 - \gamma_5)u(p)]\,[\bar{u}(p')\gamma_\rho(1 - \gamma_5)v(k')] \qquad (D.2)$$

(for the sake of brevity, polarizations are not marked explicitly in the Dirac spinors in (D.1), (D.2)). Throughout our calculations the neutrino is taken to be massless, but we will keep $m_e \neq 0$. We will also use a shorthand notation ν instead of ν_e and m instead of m_e. First let us consider the process $\nu e \to \nu e$. From (D.1) it follows easily (for an arbitrary combination of polarizations) that

$$
\begin{aligned}
|\mathcal{M}_{fi}^{(a)}|^2 &= \frac{G_F^2}{2}[\bar{u}(p')\gamma^\rho(1 - \gamma_5)u(k)]\,[\bar{u}(k)\gamma^\sigma(1 - \gamma_5)u(p')] \times \\
&\quad \times\ [\bar{u}(k')\gamma_\rho(1 - \gamma_5)u(p)]\,[\bar{u}(p)\gamma_\sigma(1 - \gamma_5)u(k')] \\
&= \frac{G_F^2}{2}\mathrm{Tr}[u(p')\bar{u}(p')\gamma^\rho(1 - \gamma_5)u(k)\bar{u}(k)\gamma^\sigma(1 - \gamma_5)] \times \\
&\quad \times\ \mathrm{Tr}[u(k')\bar{u}(k')\gamma_\rho(1 - \gamma_5)u(p)\bar{u}(p)\gamma_\sigma(1 - \gamma_5)]
\end{aligned}
$$

Summing in the last expression over polarizations (using Eq. (B.5)) we get (using also the relation $(1 - \gamma_5)^2 = 2(1 - \gamma_5)$ and other well-known properties of Dirac matrices)

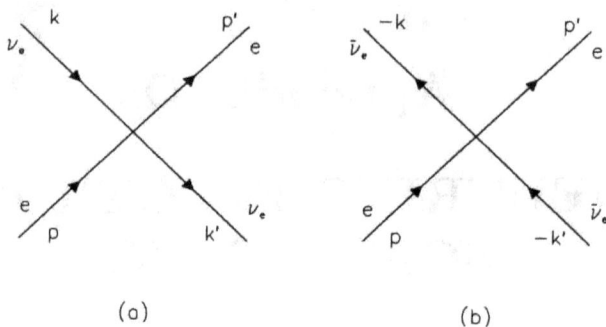

Fig. 38. Feynman diagrams corresponding to the elastic scattering processes (a) $\nu_e e \to \nu_e e$ and (b) $\bar{\nu}_e e \to \bar{\nu}_e e$ in the lowest order of perturbation expansion in a Fermi-type model.

$$\sum_{pol.} |\mathcal{M}_{fi}^{(a)}|^2 =$$

$$= 2G_F^2 \text{Tr}[(\not{p}' + m)\gamma^\rho \not{k}\gamma^\sigma (1 - \gamma_5)]\text{Tr}[\not{k}'\gamma_\rho(\not{p} + m)\gamma_\sigma(1 - \gamma_5)]$$
$$= 2G_F^2 \text{Tr}[\not{p}'\gamma^\rho \not{k}\gamma^\sigma (1 - \gamma_5)]\text{Tr}[\not{k}'\gamma_\rho\not{p}\gamma_\sigma(1 - \gamma_5)] \qquad (D.3)$$

(Notice that the terms involving m do not contribute in the last expression, since the trace of a product of an odd number of Dirac matrices vanishes.) The spinor traces in (D.3) may be evaluated most efficiently by using the following identities:

$$\text{Tr}(\not{a}\gamma^\rho\not{b}\gamma^\sigma)\text{Tr}(\not{c}\gamma_\rho\not{d}\gamma_\sigma) = 32[(a.c)(b.d) + (a.d)(b.c)]$$
$$\text{Tr}(\not{a}\gamma^\rho\not{b}\gamma^\sigma\gamma_5)\text{Tr}(\not{c}\gamma_\rho\not{d}\gamma_\sigma\gamma_5) = 32[(a.c)(b.d) - (a.d)(b.c)]$$
$$\text{Tr}(\not{a}\gamma^\rho\not{b}\gamma^\sigma)\text{Tr}(\not{c}\gamma_\rho\not{d}\gamma_\sigma\gamma_5) = 0 \qquad (D.4)$$

Let us remark that the identities (D.4) follow easily from the standard formulae (remember that we adopt the convention $\varepsilon_{0123} = +1$)

$$Tr(\gamma_\mu\gamma_\nu\gamma_\rho\gamma_\sigma) = 4(g_{\mu\nu}g_{\rho\sigma} - g_{\mu\rho}g_{\nu\sigma} + g_{\mu\sigma}g_{\nu\rho})$$
$$Tr(\gamma_\mu\gamma_\nu\gamma_\rho\gamma_\sigma\gamma_5) = 4i\varepsilon_{\mu\nu\rho\sigma}$$

Now using the formulae (D.4) in Eq. (D.3) we get the result

$$\sum_{pol.} |\mathcal{M}_{fi}^{(a)}|^2 = 128G_F^2(k.p)(k'.p')$$

which may be rewritten in terms of the Mandelstam variable s (see (A.3)) as

$$\sum_{pol.} |\mathcal{M}_{fi}^{(a)}|^2 = 32G_F^2(s - m^2)^2 \qquad (D.5)$$

For the process $\bar{\nu}e \to \bar{\nu}e$, the starting point is the expression (D.2); the corresponding calculation is completely analogous to the preceding case and it leads to the result

$$\sum_{pol.} |\mathcal{M}_{fi}^{(b)}|^2 = 32 G_F^2 (u - m^2)^2 \tag{D.6}$$

where u is the Mandelstam variable defined in (A.3) (i.e. $u = (k - p')^2$). The expressions (D.5) and (D.6) are thus related by the replacement $s \leftrightarrow u$, as was to be expected on the basis of the "crossing symmetry" (see e.g. [20], §66).

Cross sections of the considered processes may now be calculated by means of the formula (C.11) (let us recall that for an elastic scattering one always has $|\vec{p}_{c.m.}| = |\vec{p}'_{c.m.}|$). For the angular distribution of the final-state particles in the c.m. system (averaging in (D.5), (D.6) over the electron polarizations) we thus get

$$\frac{d\sigma^{(\nu e)}}{d\Omega} = \frac{G_F^2}{4\pi^2} \frac{(s - m^2)^2}{s} \tag{D.7}$$

$$\frac{d\sigma^{(\bar{\nu}e)}}{d\Omega} = \frac{G_F^2}{4\pi^2} \frac{(u - m^2)^2}{s} \tag{D.8}$$

Thus, the angular distribution of scattered particles in the process $\nu e \to \nu e$ is manifestly isotropic (in the c.m. system) according to (D.7). In order to express the right-hand side of (D.8) in terms of the scattering angle in the c.m. system we may use the relation $u = 2m^2 - s - t$ (see (A.4)) and (owing to $m_\nu = 0$)

$$t \equiv (k - k')^2 = -2kk' = -2|\vec{p}_{c.m.}|^2 (1 - \cos\vartheta)$$

Using (A.8) for the $|\vec{p}_{c.m.}|$, after a simple manipulation one gets

$$u - m^2 = -(s - m^2) \left[1 - \frac{s - m^2}{2s}(1 - \cos\vartheta) \right] \tag{D.9}$$

Of course, for $m = 0$ is (D.9) reduced to $u = -\frac{1}{2}(1 + \cos\vartheta) = -s(1 - y)$ as expected (cf. (A.6), (A.7)).

Substituting (D.9) into (D.8), we have the following result for the differential cross section of the process $\bar{\nu}e \to \bar{\nu}e$ w.r.t. scattering angle in the c.m. system:

$$\frac{d\sigma^{(\bar{\nu}e)}}{d\Omega} = \frac{G_F^2}{4\pi^2} \frac{(s - m^2)^2}{s} \left[1 - \frac{s - m^2}{2s}(1 - \cos\vartheta) \right]^2 \tag{D.10}$$

Let us now calculate the corresponding integral cross sections. The angular integration is trivial for the process $\nu e \to \nu e$; from (D.7) we get immediately

$$\sigma(\nu e \to \nu e) = \frac{G_F^2}{\pi} \frac{(s - m^2)^2}{s} \tag{D.11}$$

The integration of the differential cross section (D.10) leads to

$$\sigma(\bar{\nu}e \to \bar{\nu}e) = \frac{G_F^2}{3\pi}(s - m^2)\left[1 - (\frac{m^2}{s})^3\right] \qquad (D.12)$$

In the high-energy limit, i.e. for $s \gg m^2$, the relations (D.11), (D.12) yield approximate asymptotic formulae

$$\sigma(\nu e \to \nu e)|_{s \gg m^2} \approx \frac{G_F^2}{\pi}s \qquad (D.13)$$

$$\sigma(\bar{\nu}e \to \bar{\nu}e)|_{s \gg m^2} \approx \frac{G_F^2}{3\pi}s \qquad (D.14)$$

For completeness let us also give a simple formula for computing numerical values of the cross sections (D.13), (D.14). Since we employ a system of units in which $\hbar = c = 1$ in all relevant formulae, in order to express the cross sections in units $[\text{cm}^2]$ one has to use the conversion constant $\hbar c \doteq 0.197\text{GeVfm}$ (where $1\text{fm} = 10^{-13}\text{cm}$). Further, taking into account that $G_F \doteq 1.166 \times 10^{-5}\text{GeV}^{-2}$, then e.g. from (D.13) one gets

$$\sigma(\nu e \to \nu e) \doteq 1.7s[\text{GeV}^2] \times 10^{-38}\text{cm}^2 \qquad (D.15)$$

If one wants to express the numerical value of (D.15) as a function of the neutrino energy E_ν in the laboratory frame (i.e. in the rest system of the electron), one may use an approximate relation valid in high-energy limit (i.e. for $E_\nu \gg m$), namely

$$s \doteq 2mE_\nu \qquad (D.16)$$

Since $m \doteq 0.5\text{MeV}$, from (D.15) and (D.16) one gets

$$\sigma(\nu e \to \nu e) \doteq 1.7E_\nu[\text{GeV}] \times 10^{-41}\text{cm}^2 \qquad (D.17)$$

An unbounded growth of the cross sections (D.13), (D.14) for $s \to \infty$ means, roughly speaking, that weak interactions in a Fermi-type theory become "strong" in the high-energy limit. In this context (and for an elucidation of the term "weak interaction") it is instructive to compare numerical values of the cross sections (D.13), (D.14) with the cross section of a typical electromagnetic process (e.g. $e^-e^+ \to \mu^-\mu^+$) for various energies. Let us recall that for $s \gg m_\mu^2$ one has, in the tree approximation (i.e. in the 2nd order of perturbation expansion in QED), the approximate formula

$$\sigma(e^-e^+ \to \mu^-\mu^+) \doteq \frac{4\pi\alpha^2}{3s} \doteq \frac{86.8}{s(\text{GeV}^2)}\text{nb} \qquad (D.18)$$

where $1\text{nb} (= 1 \text{ nanobarn}) = 10^{-33} \text{ cm}^2$ (α is the fine-structure constant, $\alpha \doteq 1/137$).

Appendix E

JACOB-WICK EXPANSION

In this appendix we present some basic relations and formulae concerning the expansion of a relativistic scattering amplitude (given in momentum and helicity representation) into partial waves (characterized by values of the total angular momentum), i.e. the so-called Jacob-Wick expansion [19]. Within the framework of such a formalism we then discuss the condition of unitarity of the S-matrix. A more detailed exposition and a derivation of the Jacob-Wick expansion may be found either in the original paper [19] or in the textbooks [20], [21]. A very useful survey of this method is also contained in an appendix of the paper [22].

First we will consider a process of elastic scattering of particles *1, 2*; throughout our discussion we are working in the centre-of-mass system. The initial and final states of both particles are characterized by their momenta (they are plane waves) and helicities. The axis 3 of the coordinate frame will be identified with the direction of an initial-state particle momentum. For the scattering amplitude normalized so that its square is just equal to the differential cross section, i.e.

$$\frac{d\sigma}{d\Omega} = |f|^2 \qquad (\text{E.1})$$

one may write the partial-wave expansion (Jacob-Wick expansion [19])

$$f_{h'h}(s,\Omega) = \sum_j (2j+1) f_{h'h}^{(j)}(s) \mathcal{D}_{\lambda'\lambda}^{(j)}(\Omega) \qquad (\text{E.2})$$

where $h \equiv (h_1, h_2)$, and $h' \equiv (h'_1, h'_2)$ are the initial and final helicities, $\Omega \equiv (\vartheta, \varphi)$ defines a direction of the momentum of scattered particles and $\mathcal{D}_{\lambda'\lambda}^{(j)}(\Omega)$ are Wigner functions (known also from the theory of angular momentum as the matrix elements of finite rotations - see e.g. [23]). The indices λ, λ' are given by

$$\lambda = h_1 - h_2, \quad \lambda' = h'_1 - h'_2$$

Some basic properties of the Wigner $\mathcal{D}$-functions are summarized in Appendix F. A coefficient $f^{(j)}$ in the expansion (E.2) is the amplitude of the partial wave corresponding to the total angular momentum j. The sum in (E.2) runs over all non-negative integer or half-integer values of the j depending on whether the set of particles *1, 2* contains an even or an odd number of fermions. The amplitudes of partial waves have the form

$$f_{h'h}^{(j)}(s) = \frac{1}{2i|\vec{p}|}(S_{h'h}^{(j)} - 1) \tag{E.3}$$

where $\vec{p}$ is the momentum of colliding particles in the c.m. system (for elastic scattering we of course have $|\vec{p}| = |\vec{p'}|$) and $S_{h'h}^{(j)}$ is S-matrix element for scattering in a state with total angular momentum j and for given initial and final helicities h and h'). The essential point is that $S_{h'h}^{(j)}$ belongs to a unitary matrix. This immediately implies an important bound for the partial-wave amplitude $f^{(j)}(s)$ (here and in what follows we usually omit the indices h, h')

$$|f^{(j)}(s)| \leq \frac{1}{|\vec{p}|} \tag{E.4}$$

Let us recall that $|\vec{p}|$ can be expressed in terms of s as

$$|\vec{p}| = \frac{\lambda^{\frac{1}{2}}(s, m_1^2, m_2^2)}{2s^{\frac{1}{2}}}$$

(see (A.8)). The expansion (E.2) may be rewritten for the Lorentz-invariant scattering amplitude $\mathcal{M}$ which we usually employ in our calculations (which is defined directly as a contribution of Feynman diagrams). Indeed, comparing the formulae for scattering cross section (E.1) and (C.11) (equating phases of f and $\mathcal{M}$) in general, i.e. including an inelastic scattering case, where $|\vec{p}| \neq |\vec{p'}|$ one gets

$$\mathcal{M} = 8\pi \, s^{\frac{1}{2}} \left(\frac{|\vec{p}|}{|\vec{p'}|}\right)^{\frac{1}{2}} f \tag{E.5}$$

The partial-wave expansion for the amplitude $\mathcal{M}$ may be written as

$$\mathcal{M}_{h'h}(s, \Omega) = 16\pi \sum_j (2j + 1)\mathcal{M}_{h'h}^{(j)}(s)\mathcal{D}_{\lambda'\lambda}^{(j)}(\Omega) \tag{E.6}$$

(the coefficient 16π in (E.6) is chosen conventionally for a convenient normalization of the amplitudes $\mathcal{M}^{(j)}$ - see below, the relation (E.12)). From an orthogonality relation for the Wigner $\mathcal{D}$-functions (see (F.6) in Appendix F) we obtain for partial-wave amplitudes in (E.6) a general formula

$$\mathcal{M}^{(j)}(s) = \frac{1}{16\pi} \int \mathcal{M}(s, \Omega)\mathcal{D}_{\lambda'\lambda}^{*(j)}(\Omega)\frac{d\Omega}{4\pi} \tag{E.7}$$

In the particular case where $\lambda' = \lambda = 0$ (i.e. for $h_1 = h_2$, $h_1' = h_2'$) the $\mathcal{D}$-functions are reduced to Legendre polynomials (see (F.4)) and the formula (E.7) then becomes

$$\mathcal{M}^{(j)}(s) = \frac{1}{32\pi} \int_{-1}^{1} \mathcal{M}(s, \vartheta)P_j(\cos\vartheta)d(\cos\vartheta) \tag{E.8}$$

In an elastic scattering case, the relation (E.5) simplifies to

$$\mathcal{M} = 8\pi\sqrt{s}\,f \tag{E.9}$$

From (E.3) and (E.9) we thus get

$$\mathcal{M}^{(j)}(s) = \frac{\sqrt{s}}{4i|\vec{p}|}(S^{(j)} - 1) \tag{E.10}$$

and unitarity of the matrix $S^{(j)}$ then yields the bound

$$|\mathcal{M}^{(j)}(s)| \le \frac{\sqrt{s}}{2|\vec{p}|} \tag{E.11}$$

In high-energy limit or for massless particles one has $|\vec{p}| \approx \frac{1}{2}\sqrt{s}$ and instead of (E.11) we may write a simpler inequality

$$|\mathcal{M}^{(j)}(s)| \le 1 \tag{E.12}$$

For an inelastic process $1 + 2 \to 3 + 4$ one may also write a partial-wave expansion in the form (E.2) or (E.6); however, in such a case only the purely non-diagonal S-matrix elements are involved. Instead of (E.3) and (E.9) (cf. also [23], where the case of spin-zero particles is discussed) we then have

$$f_{inel.}^{(j)}(s) = \frac{1}{2i|\vec{p}|^{\frac{1}{2}}|\vec{p}'|^{\frac{1}{2}}}S_{inel.}^{(j)} \tag{E.13}$$

or

$$\mathcal{M}_{inel.}^{(j)}(s) = \frac{s^{\frac{1}{2}}}{4i|\vec{p}'|}S_{inel.}^{(j)} \tag{E.14}$$

where the symbol $S_{inel.}^{(j)}$ again represents collectively elements of the relevant unitary matrix and the index "inel." denotes the inelastic channel $1 + 2 \to 3 + 4$. In high-energy limit, the relation (E.14) implies the bound

$$|\mathcal{M}_{inel.}^{(j)}| \le \frac{1}{2} \tag{E.15}$$

The constraints for partial-wave amplitudes following from S-matrix unitarity can also be easily converted into inequalities for partial cross sections (i.e. for cross sections corresponding to the individual partial waves). From (C.11), (E.6) and using the orthogonality relation (F.6) for the $\mathcal{D}$-functions in the expansion (E.6), after performing the angular integration (for a given set of the initial and final helicities) we get

$$\sigma(s) = \sum_j \sigma^{(j)}(s) \tag{E.16}$$

where

$$\sigma^{(j)}(s) = \frac{16\pi}{s}(2j+1)|\mathcal{M}^{(j)}(s)|^2 \qquad (E.17)$$

In the case of elastic scattering, the inequality (E.11) then implies a bound for the partial cross sections (E.17), namely

$$\sigma^{(j)}(s) \le (2j+1)\frac{4\pi}{|\vec{p}|^2} \qquad (E.18)$$

which in high-energy limit becomes

$$\sigma^{(j)}(s) \le (2j+1)\frac{16\pi}{s} \qquad (E.19)$$

In the case of an inelastic process it is easy to derive analogous inequalities; in high-energy limit (or for massless particles) from (E.17) and (E.15) one gets

$$\sigma^{(j)}_{inel.}(s) \le (2j+1)\frac{4\pi}{s} \qquad (E.20)$$

Appendix F

WIGNER $\mathcal{D}$-FUNCTIONS

In this appendix we summarize some important properties of the Wigner $\mathcal{D}$-functions which enter the Jacob-Wick expansion described in Appendix E. A more detailed review may be found e.g. in [20] or [23].

In what follows, the symbol Ω denotes, as ever, a pair of spherical angles defining a direction in the 3-dimensional space. Wigner $\mathcal{D}$-function appearing in the expansion (E.2) or (E.6) is defined by

$$\mathcal{D}^{(j)}_{m'm}(\Omega) = e^{im\varphi} d^{(j)}_{m'm}(\vartheta) \tag{F.1}$$

Indices m, m' may only take on values $-j, -j+1, ..., j-1, j$, and the functions $d^{(j)}_{m'm}(\vartheta)$ are given by the general formula

$$d^{(j)}_{m'm}(\vartheta) = (-1)^{j-m'} \left[\frac{(j+m')!}{2^{2j}(j-m')!(j+m)!(j-m)!} \right]^{\frac{1}{2}} \times$$

$$\times (1+\xi)^{\frac{m'+m}{2}} (1-\xi)^{-\frac{m'-m}{2}} \left(\frac{d}{d\xi} \right)^{j-m'} \left[(1+\xi)^{j+m} (1-\xi)^{j-m} \right] \tag{F.2}$$

where $\xi = \cos\vartheta$.

Some special properties of $d^{(j)}_{m'm}(\vartheta)$:

$$\begin{aligned} d^{(j)}_{m'm}(-\vartheta) &= d^{(j)}_{mm'}(\vartheta) \\ d^{(j)}_{m'm}(\vartheta) &= d^{(j)}_{-m-m'}(\vartheta) \\ d^{(j)}_{m'm}(0) &= \delta_{m'm} \end{aligned} \tag{F.3}$$

When $m = m' = 0$, for an arbitrary integer $l \geq 0$

$$\mathcal{D}^{(l)}_{00}(\Omega) = P_l(\cos\vartheta) \tag{F.4}$$

where P_l is Legendre polynomial.

Examples of explicit form of the functions $d^{(j)}_{m'm}(\vartheta)$ for $j = 1$:

$$d^{(1)}_{11} = d^{(1)}_{-1-1} = \frac{1}{2}(1+\cos\vartheta)$$

$$d_{00}^{(1)} = \cos \vartheta \tag{F.5}$$

$$d_{1-1}^{(1)} = d_{-11} = \frac{1}{2}(1 - \cos \vartheta)$$

$$d_{10}^{(1)} = -d_{01}^{(1)} = d_{0-1}^{(1)} = -d_{-10}^{(1)} = \frac{1}{\sqrt{2}} \sin \vartheta$$

An orthogonality relation:

$$\int \mathcal{D}_{m_1' m_1}^{*(j_1)}(\Omega) \mathcal{D}_{m_2' m_2}^{(j_2)}(\Omega) \frac{d\Omega}{4\pi} = \frac{1}{2j_1 + 1} \delta_{j_1 j_2} \delta_{m_1 m_2} \tag{F.6}$$

Appendix G

INDEX OF FEYNMAN DIAGRAM

In this appendix we derive, for completeness, a standard formula for the "index" (or "superficial degree of divergence") of an arbitrary Feynman diagram within the framework of a general model of quantum field theory described by a polynomial Lagrangian (see also [21]). We discuss separately interactions involving a spin-1 boson field with a non-zero mass (M) which is not treated in sufficient detail in [21]: In such a case, if we use the canonical propagator of the massive vector field which behaves in the ultraviolet (UV) region like a constant $\simeq M^{-2}$ (see (H.45) in Appendix H), then the standard formula for the index of a Feynman graph [21] should be modified in a simple way, as we will show in the sequel (see also [25], [26]).

First we are going to discuss a "standard" case where all boson propagators (in momentum representation) behave in UV region as k^{-2}. The contribution of a Feynman graph involving L closed loops (i.e. L independent momenta of internal lines) may be written as

$$\mathcal{M}(G) = \int d^4 k_1 ... d^4 k_L \ \ \mathcal{J}(k_1, ..., k_L; \ p_{ext.}) \tag{G.1}$$

where $k_1, ..., k_L$ are relevant internal (loop) momenta and the symbol $p_{ext.}$ denotes collectively external momenta; in (G.1) we have neglected a possible dependence on masses of particles corresponding to the internal lines (i.e. propagators) since a non-zero mass in a propagator obviously does not influence the convergence properties of the integral (G.1) in the UV region $k_i \rightarrow \infty$, $i = 1, ..., k$. The integrand in (G.1) is thus a homogeneous function of the variables $k_1, ..., k_L$ in the UV region. We then define the index of the graph G as the degree of homogeneity of the complete expression behind the integration sign in (G.1) (i.e. including $d^4 k_1 ... d^4 k_L$) and denote it as $\omega(G)$; this means that when rescaling the loop momenta according to

$$k_i \rightarrow \lambda k_i, \ \ i = 1, ..., L \tag{G.2}$$

the expression behind the integration sign in (G.1) (where all the masses are neglected) is multiplied by the factor $\lambda^{\omega(G)}$. It is easy to realize that $\omega(G) < 0$ corresponds to a convergent integral (G.1) (which however may contain UV-divergent subgraphs) and for (superficially) UV-divergent graphs one has $\omega(G) \geq 0$ (such an UV divergence is logarithmic for $\omega(G) = 0$, linear for $\omega(G) = 1$, quadratic for $\omega(G) = 2$ etc.). Let us stress that in such a simple estimate of the degree of divergence of a Feynman graph based on a straightforward power counting in (G.1) we have of course ignored any subtle details of the considered diagram which in particular cases may cause an "accidental" cancellation of some of the potential UV divergences. A terminological remark is perhaps also in order here. In the literature, the $\omega(G)$ is often called "superficial degree of divergence" or "overall degree of divergence" of a graph. We employ here the term "index" (which is frequently used e.g. in Russian literature) mostly for the sake of brevity and terminological simplicity, taking into account that later we will also introduce the notion of an "index" or "effective index" of an interaction vertex.

In order to calculate the $\omega(G)$ one has to realize that under the scaling transformation (G.2) in the UV region, each fermion propagator is multiplied by a factor λ^{-1}, each boson propagator yields (according to our assumption) a factor λ^{-2} and a derivative from the interaction Lagrangian (acting on an internal line) gives a factor of λ; finally, the volume element in (G.1) contributes a factor λ^{4L}. Putting this together we get

$$\omega(G) = 4L - I_F - 2I_B + \sum_v \delta_v \qquad (G.3)$$

where I_F is the number of internal fermion lines of the considered graph, I_B is the number of internal boson lines and δ_v is the number of derivatives from interaction Lagrangian acting in a vertex v on the internal lines and the sum in (G.3) runs over all vertices of the graph G. The number of closed loops L may easily be expressed in terms of the total number of internal lines (I) and total number of vertices (V):

$$L = I - V + 1 \qquad (G.4)$$

Of course, one has $I = I_F + I_B$ and Eq. (G.3) may thus be rewritten as

$$\omega(G) - 4 = 3I_F + 2I_B - 4V + \sum_v \delta_v \qquad (G.5)$$

The number of internal fermion or boson lines may be expressed as

$$I_F = \frac{1}{2}\sum_v f_v$$

$$I_B = \frac{1}{2}\sum_v b_v \qquad (G.6)$$

where f_v or b_v is the number of internal fermion or boson lines attached to the vertex v. Further, one has

$$f_v = n_{F;v} - E_{F;v}$$
$$b_v = n_{B;v} - E_{B;v}$$
$$\delta_v = n_{D;v} - E_{D;v} \qquad (G.7)$$

where $E_{F;v}$ is the number of external fermion lines attached to the vertex v, the $E_{B;v}$ has the same meaning for boson lines and $E_{D;v}$ denotes the number of derivatives from the interaction term corresponding to the vertex v which act on external lines. Similarly, the symbols $n_{F;v}$ and $n_{B;v}$ in (G.7) denote the total numbers of fermion and boson lines attached to the vertex v (i.e. the total numbers of fermion and boson fields occurring in the corresponding term of the interaction Lagrangian) and $n_{D;v}$ is the total number of derivatives in the corresponding interaction term. Using (G.6) and (G.7), the relation (G.5) may be recast as

$$\omega(G) - 4 = \sum_v (\omega_v - 4) - (\frac{3}{2}E_F + E_B + \delta) \qquad \text{(G.8)}$$

where we have introduced the notation

$$\omega_v = \frac{3}{2}n_{F;v} + n_{B;v} + n_{D;v} \qquad \text{(G.9)}$$

and

$$
\begin{aligned}
E_F &= \sum_v E_{F;v} \\
E_B &= \sum_v E_{B;v} \\
\delta &= \sum_v E_{D;v} \qquad \text{(G.10)}
\end{aligned}
$$

The E_F (E_B) is thus the total number of external fermion (boson) lines of the considered Feynman diagram and δ is the total power of external momenta factorized in the contribution of the graph as a result of the action of derivatives from interaction terms on the external lines. The number ω_v defined by eq.(G.9) is usually called the index of the vertex v and it characterizes a corresponding term in the interaction Lagrangian. The values of ω_v for individual interaction terms (i.e. for individual vertices of diagrams) in a sense determine, according to (G.8), the structure of UV divergences of Feynman graphs in a given model of quantum field theory and indicate thus renormalizability or non-renormalizability of the perturbation expansion: If there is $\omega_v > 4$ for at least one interaction vertex in the considered model, then on the basis of (G.8) one may in general expect an infinite number of types of UV divergences (i.e. there is an infinite number of combinations of E_F and E_B for which one may get a UV-divergent graph in a sufficiently high order of perturbation expansion) and such a field theory model is then "suspect" of being non-renormalizable (however, a special additional mechanism may operate, cancelling the offending UV-divergences so that the perturbation expansion may turn out to be renormalizable despite an "unfavourable" power-counting result). If for any vertex one has $\omega_v \leq 4$, there may be only a finite number of types of UV-divergent graphs (here one should emphasize that in (G.8) one of course has $E_F \geq 0$, $E_B \geq 0$ and $\delta \geq 0$) and the perturbation expansion is thus renormalizable by means of a finite number of counterterms.

In this connection, it is also useful to notice that the value of ω_v given by (G.9) is equal to the dimension of the corresponding interaction term $\mathcal{L}_{int}^{(v)}$ (i.e. of the corresponding monomial in relevant fields, without a coupling constant) in units of an arbitrary mass M: Indeed, the dimension of a fermion field (i.e. the corresponding power of M) is equal to $\frac{3}{2}$ and the dimension of any boson field is equal to 1, as one may easily find from the corresponding free Lagrangians; the dimension of a derivative is of course equal to 1. The formula (G.9) may thus be recast as

$$\omega_v = n_{F;v}\mathrm{dim}\psi + n_{B;v}\mathrm{dim}B + n_{D;v}\,\mathrm{dim}\partial \tag{G.11}$$

and the right-hand side of the last expression is equal to $\mathrm{dim}\mathcal{L}_{int}^{(v)}$. (The symbol $\mathrm{dim}X$ has of course the same meaning as the notation $[X]$ used for a canonical dimension in other places of this text.) Let us remark that the formula (G.11) is generally valid in an n-dimensional space for $n \neq 4$, if we use the appropriate values of $\mathrm{dim}\psi$ and $\mathrm{dim}B$; such a generalization of the relation (G.11) is left to the interested reader as an instructive exercise.

Let us now consider a model of quantum field theory where all the boson fields have spin 1 and a non-zero mass and take the corresponding propagators to have the canonical form (H.45) (an example of such a model is the theory of weak interactions with a charged IVB described in Chapter 3).In such a case the boson propagators behave in the UV region as a non-zero constant and the preceding calculation of the index of a Feynman graph is modified in a simple way: In the basic formula (G.3) one has to replace the term $-2I_B$ by zero. Further steps in the computation of $\omega(G)$ are not changed and the above-mentioned modification of eq. (G.3) thus implies that instead of the previous results (G.8), (G.9) now one gets

$$\omega(G) - 4 = \sum_v (\omega_v^{eff.} - 4) - (\frac{3}{2}E_F + 2E_B + \delta) \tag{G.12}$$

where we have denoted

$$\omega_v^{eff.} = \frac{3}{2}n_{F;v} + 2n_{B;v} + n_{D;v} \tag{G.13}$$

All preceding considerations may easily be generalized to the case of a field theory model involving boson fields both of the type 1 (with the propagator $\simeq k^{-2}$ in the UV region) and of the type 2 (with the propagator $\simeq const.$ in the UV region): In such a case, the formulae (G.9) and (G.13) are combined to

$$\omega_v^{eff.} = \frac{3}{2}n_{F;v} + n_{B;v}^{(1)} + 2n_{B;v}^{(2)} + n_{D;v} \tag{G.14}$$

where $n_{B;v}^{(1)}$ or $n_{B;v}^{(2)}$ is the number of the type-1 or type-2 boson lines attached to the vertex v and the second term in (G.8) or (G.12) is modified analogously.

The number $\omega_v^{eff.}$ appearing in (G.12), (G.13) or (G.14) will be called an "effective index" of the interaction vertex v. The adjective "effective" should reflect the fact that the formulae (G.13) or (G.14) describe a structure of the UV divergences assuming that one employs the canonical propagator (H.45) for massive vector fields; the value of $\omega_v^{eff.}$ thus provides an information on potential UV divergences arising as a combined effect of the structure of the corresponding interaction term and a "bad" high-energy behaviour of the propagator (H.45). It is in order to remark here that the above-mentioned canonical description of a massive vector field is not always mandatory; generally speaking, one may use a formalism involving a type-1 vector propagator and an auxiliary unphysical spin-zero field (see e.g. [21], paragraph 3.2.3). Internal consistency of such a formalism (i.e. the fact that the unphysical auxiliary field does not influence physical quantities) must in each case be verified separately. Thus, e.g., in spinor electrodynamics with a massive photon, such a formalism is internally consistent and the same is true for non-abelian gauge theories with the Higgs mechanism; these theories are renormalizable (although the relevant effective indices $\omega_v^{eff.}$ calculated from Eq. (G.13) or (G.14) suggest non-renormalizability of the perturbation expansions). In both of these cases, a gauge symmetry (abelian in QED case) is essential. However, the above-mentioned alternative formalism for the description of a massive vector field cannot be used consistently e.g. in the model of weak interactions with a charged IVB described in Chapter 3 or in the electrodynamics of charged vector bosons (Chapter 4). The difficulty is that in both cases one gets a non-unitary S-matrix at higher orders of perturbation expansion (see e.g.[26], [29]). Within the framework of the canonical formalism, both these models are non-renormalizable, in accordance with an estimate based on the formula (G.13) or (G.14).

In any case one may say that a value of the effective index $\omega_v^{eff.} > 4$ in models involving interactions of massive vector bosons is signalling potential problems with UV divergences in high orders of perturbation expansion which, however, may in fact sometimes be suppressed by means of more subtle special mechanisms. A physically relevant example of such an interesting situation is the standard model of electroweak interactions described in Chapter 5. All the interaction terms of course satisfy the condition $\dim \mathcal{L}_{int}^{(v)} \leq 4$.

From what we have said up to now it should be clear that it makes sense to distinguish between the effective index $\omega_v^{eff.}$ defined by Eq. (G.13) or (G.14) and the index ω_v which may always be defined as the dimension of the corresponding interaction term (cf. the formulae (G.9) and (G.11)). Of course, in some particular cases the equality $\omega_v = \omega_v^{eff.}$ may hold trivially (as e.g. in a Fermi-type theory, i.e. in a model of direct four-fermion interaction).

Appendix H

MASSIVE VECTOR FIELD

In this appendix we summarize some basic properties of a massive vector field, i.e. the field corresponding to massive spin -1 particles and we derive here some important relations which are used frequently in the main text in the description of processes involving intermediate vector bosons. Further details may be found e.g. in the textbooks [21] (§3.2.3), [36] (§2.8 and §4.5).

Let us first consider the relativistic wave equation for a free particle with spin 1 and a non-zero mass which was originally formulated by Proca (see e.g. [36], [37]):

$$\partial_\mu F^{\mu\nu} + m^2 B^\nu = 0 \qquad (H.1)$$

where

$$F^{\mu\nu} = \partial^\mu B^\nu - \partial^\nu B^\mu \qquad (H.2)$$

Equations (H.1), (H.2) represent, in a sense, a straightforward generalization of Maxwell equations (which correspond to massless photons). The corresponding one-particle wave function is described here by four (in general complex) functions of space-time coordinates $B^\mu(x)$ ($\mu = 0, 1, 2, 3$) which are components of a four-vector with respect to Lorentz transformations and the parameter $m \neq 0$ in (H.1) has dimension of a mass. (The presence of a mass term of course means that Proca equations are not invariant under gauge transformations.)

Substituting (H.2) into (H.1) one gets

$$(\Box + m^2)B^\nu - \partial^\nu(\partial_\mu B^\mu) = 0 \qquad (H.3)$$

Acting on eq. (H.3) with ∂_ν (i.e. calculating the four-divergence of (H.3)) then on the left-hand side only the expression $m^2 \partial_\nu B^\nu$ remains and thus we get immediately

$$\partial_\nu B^\nu = 0, \qquad (H.4)$$

i.e. a "Lorentz condition" follows automatically from Proca equations (H.1), (H.2). The essential point in derivation of Eq. (H.4) is the fact that $m \neq 0$, i.e. that the original equation (H.1) contains a mass term. (When $m = 0$ we get only a trivial identity by means of the same procedure; this corresponds to the well-known fact that the Lorentz condition does not follow from Maxwell equations but rather represents an appropriately chosen subsidiary condition.)

The result (H.4) means that Proca equation (H.3) for the four-vector B^μ is equivalent to the pair of equations

$$(\Box + m^2)B^\mu = 0, \qquad \partial^\mu B_\mu = 0 \tag{H.5}$$

That is, individual components of the wave function B^μ satisfy the Klein-Gordon equation (and thus indeed describe a particle with mass m) but they are not independent since the four-divergence of the B^μ vanishes. The equations (H.4) physically mean that the number of independent components of the wave function is thus reduced (in a covariant manner) to three, which correspond to a spin-1 particle. The independent components are conveniently chosen to be B^j, $j = 1, 2, 3$ and B^0 may then be expressed in terms of B^j using (H.4). (Let us remark that (H.4) in fact represents the only conceivable Lorentz-covariant condition linear in B^μ which eliminates just one degree of freedom in the considered four-component wave function.)

We will now examine solutions of equations (H.1), (H.2) or the equivalent equations (H.5), corresponding to a given momentum $\vec{k}$. Such a plane-wave solution may be written as

$$B_\mu(x) = N(k)\varepsilon_\mu(k)e^{-ikx} \tag{H.6}$$

where $k^\mu = (k_0, \vec{k})$, and from the Klein-Gordon equation in (H.5) it immediately follows that

$$k^2 = k_o^2 - \vec{k}^2 = m^2 \tag{H.7}$$

i.e. k is the four-momentum of a particle with mass m. (A remark: Here and in what follows, if we write components of a four-vector without denoting them explicitly we always have in mind upper Lorentz indices, i.e. the contravariant components.) The $N(k)$ in (H.6) is a normalization factor whose specific value is inessential at present and $\varepsilon_\mu(k)$ represents the wave function in momentum space; in this sense it is e.g. a direct analogy of the functions $u(k)$, $v(k)$ in Dirac plane waves (cf. Appendix B). At the same time, the $\varepsilon_\mu(k)$ may be interpreted (similarly to the case of solutions of Maxwell equations) as a polarization vector corresponding to the plane wave (H.6). Such a dual role of the four-vector $\varepsilon_\mu(k)$ is of course specific for the description of a spin-1 particle. (In what follows we will also clarify a connection between polarization and helicity for plane-wave solutions of the type (H.6).) The second equation in (H.5) immediately yields

$$k.\varepsilon(k) = 0 \tag{H.8}$$

where $k.\varepsilon(k) = k^\mu \varepsilon_\mu(k)$. In order to find all linearly independent solutions of eq. (H.8) it is instructive to consider first the corresponding solutions in the rest frame of the vector particle, i.e. for $k = k^{(0)} = (m, 0, 0, 0)$. Eq. (H.8) then implies $\varepsilon_0(k^{(0)}) = 0$; the space components $\varepsilon_j(k^{(0)})$, $j = 1, 2, 3$ may be arbitrary. There are 3 linearly independent (in general complex) three-dimensional vectors $\vec{\varepsilon}^{\,(1)}$, $\vec{\varepsilon}^{\,(2)}$, $\vec{\varepsilon}^{\,(3)}$ which may be chosen to be orthogonal, i.e. satisfying conditions

$$\vec{\varepsilon}^{\,(\lambda)}.\vec{\varepsilon}^{\,(\lambda')*} = \delta_{\lambda\lambda'} \tag{H.9}$$

for $\lambda, \lambda' = 1, 2, 3$. In the rest frame one may thus write 3 linearly independent solutions of eq. (H.8)

$$\varepsilon^{(\lambda)} = (0, \vec{\varepsilon}^{\,(\lambda)}), \qquad \lambda = 1, 2, 3 \tag{H.10}$$

which just correspond to three possible spin states of a massive vector particle. An obvious explicit example of a solution of the type (H.10) is the triplet of real vectors

$$\begin{aligned}
\varepsilon^{(1)} &= (0, 1, 0, 0) \\
\varepsilon^{(2)} &= (0, 0, 1, 0) \\
\varepsilon^{(3)} &= (0, 0, 0, 1)
\end{aligned} \tag{H.11}$$

It is useful to notice that the conditions (H.9) may be rewritten in terms of Lorentz-invariant scalar product for the four-component objects (H.10) as

$$\varepsilon^{(\lambda)}.\varepsilon^{(\lambda')*} = -\delta_{\lambda\lambda'} \tag{H.12}$$

If we require that the $\varepsilon^{(\lambda)}$ in (H.10) transform as four-vectors, a triplet of linearly independent solutions of eq. (H.8) for an arbitrary k $(k^2 = m^2)$ may be obtained from (H.10) by means of the corresponding Lorentz transformation. Denoting three linearly independent solutions of eq. (H.8) as $\varepsilon(k, \lambda)$ (where again $\lambda = 1, 2, 3$), the normalization condition (H.12) imposed in the rest frame then also implies that

$$\varepsilon(k, \lambda).\varepsilon^*(k, \lambda') = -\delta_{\lambda\lambda'} \tag{H.13}$$

for $\lambda, \lambda' = 1, 2, 3$. Vectors $\varepsilon(k, \lambda)$ for a given momentum $\vec{k}$ can easily be found directly from eq. (H.8), without performing the above-mentioned Lorentz transformation. To this end, one may choose 3 real vectors $\vec{\varepsilon}(k, \lambda)$, $\lambda = 1, 2, 3$ such that the first two are mutually orthogonal and also orthogonal to $\vec{k}$, and the $\vec{\varepsilon}(k, 3)$ is directed along the $\vec{k}$, i.e.

$$\vec{\varepsilon}(k, 3) = a\frac{\vec{k}}{|\vec{k}|} \tag{H.14}$$

where $a > 0$. A solution of eq. (H.8) may be then written as

$$\begin{aligned}
\varepsilon(k, 1) &= (0, \vec{\varepsilon}(k, 1)) \\
\varepsilon(k, 2) &= (0, \vec{\varepsilon}(k, 2)) \\
\varepsilon(k, 3) &= \left(a\frac{|\vec{k}|}{k_0}, \, a\frac{\vec{k}}{|\vec{k}|}\right)
\end{aligned} \tag{H.15}$$

The normalization condition (H.13) is satisfied if we take the $\vec{\varepsilon}\,(k,1)$ and $\vec{\varepsilon}\,(k,2)$ to be unit vectors and in the expression for $\varepsilon(k,3)$ we set $a = k_0/m$. Vectors $\varepsilon(k,\lambda)$ thus correspond to (linear) transverse polarizations for $\lambda = 1,2$ and longitudinal polarization for $\lambda = 3$. In the following we will employ the usual symbol $\varepsilon_L(k)$ for the longitudinal polarization; according to the preceding discussion, its components are given by

$$\varepsilon_L^\mu(k) = \varepsilon^\mu(k,3) = \left(\frac{|\vec{k}|}{m}, \frac{k_0}{m} \frac{\vec{k}}{|\vec{k}|} \right) \tag{H.16}$$

It is perhaps in order to emphasize that the existence of *three* nontrivial polarization vectors, i.e. of three space-like four-vectors satisfying (H.8) is obviously related to non-zero rest mass of the considered vector particle; it can best be seen from the discussion of the corresponding solutions in the rest frame, whose very existence is guaranteed by the fact that $m \neq 0$. It is easy to prove that for a massless particle there is no space-like vector satisfying (H.8) which would correspond to longitudinal polarization.

For completeness we will now clarify a connection of the polarization vectors (H.15) with states characterized by a definite helicity. Orientations of the unit vectors $\vec{\varepsilon}\,(k,1)$, $\vec{\varepsilon}\,(k,2)$ may be chosen such that

$$\vec{n} \times \vec{\varepsilon}\,(k,1) = \vec{\varepsilon}\,(k,2) \tag{H.17}$$

where $\vec{n} = \vec{k}/|\vec{k}|$ is the unit vector along the direction of $\vec{k}$. From (H.17) then it also immediately follows that

$$\vec{n} \times \vec{\varepsilon}\,(k,2) = -\vec{\varepsilon}\,(k,1) \tag{H.18}$$

The relevant hermitean 3×3 matrices representing spin components are generators of rotations in three-dimensional space around the corresponding coordinate axes, i.e.

$$S_1 = \begin{pmatrix} 0 & 0 & 0 \\ 0 & 0 & -i \\ 0 & i & 0 \end{pmatrix}, \quad S_2 = \begin{pmatrix} 0 & 0 & i \\ 0 & 0 & 0 \\ -i & 0 & 0 \end{pmatrix}, \quad S_3 = \begin{pmatrix} 0 & -i & 0 \\ i & 0 & 0 \\ 0 & 0 & 0 \end{pmatrix} \tag{H.19}$$

The helicity operator for a spin-1 particle carrying a momentum $\vec{k}$ is thus represented by the matrix

$$\hat{h}(\vec{k}) = \vec{n}.\vec{S} = i \begin{pmatrix} 0 & -n_3 & n_2 \\ n_3 & 0 & -n_1 \\ -n_2 & n_1 & 0 \end{pmatrix} \tag{H.20}$$

Acting with (H.20) on an arbitrary vector $\vec{\varepsilon}$ (viewed for convenience as a one-column matrix) it is straightforward to derive the formula

$$\hat{h}(\vec{k})\vec{\varepsilon} = i(\vec{n} \times \vec{\varepsilon}) \tag{H.21}$$

Using (H.21), (H.17) and (H.18) we then get for the polarization vectors in (H.15) or (H.16)

$$\hat{h}(\vec{k})\vec{\varepsilon}\,(k,1) = i\vec{\varepsilon}\,(k,2)$$
$$\hat{h}(\vec{k})\vec{\varepsilon}\,(k,2) = -i\vec{\varepsilon}\,(k,1) \tag{H.22}$$
$$\hat{h}(\vec{k})\vec{\varepsilon}_L(k) = 0$$

If we define complex vectors

$$\vec{\varepsilon}\,(k,+) = \frac{1}{\sqrt{2}}[\vec{\varepsilon}\,(k,1) + i\vec{\varepsilon}\,(k,2)]$$
$$\vec{\varepsilon}\,(k,-) = \frac{1}{\sqrt{2}}[\vec{\varepsilon}\,(k,1) - i\vec{\varepsilon}\,(k,2)] \tag{H.23}$$

(passing thus from linear to circular polarizations), the relations (H.22), (H.23) immediately yield

$$\hat{h}(\vec{k})\vec{\varepsilon}\,(k,+) = \vec{\varepsilon}\,(k,+)$$
$$\hat{h}(\vec{k})\vec{\varepsilon}\,(k,-) = -\vec{\varepsilon}\,(k,-) \tag{H.24}$$

Thus, (H.24) together with the last equation in (H.22) make it clear that the vectors $\vec{\varepsilon}\,(k,\pm)$ and $\vec{\varepsilon}_L(k)$ represent states with helicities ± 1 and 0.

Now we are going to derive an important relation concerning the asymptotic behaviour of components of the vector of longitudinal polarization in high-energy limit (i.e. for $|\vec{k}| \gg m$), which reads

$$\varepsilon_L^\mu(k) = \frac{1}{m}k^\mu + O\left(\frac{m}{k_o}\right) \tag{H.25}$$

The proof of (H.25) is easy. Using (H.16) one gets for the difference of the four-vectors $\varepsilon_L(k)$ and k/m first

$$\varepsilon_L^\mu(k) - \frac{1}{m}k^\mu = \left(\frac{|\vec{k}| - k_0}{m}, \frac{k_0 - |\vec{k}|}{m}\frac{\vec{k}}{|\vec{k}|}\right) \tag{H.26}$$

However,

$$\frac{k_0 - |\vec{k}|}{m} = \frac{1}{m}\frac{k_0^2 - |\vec{k}|^2}{k_0 + |\vec{k}|} = \frac{m}{k_0 + |\vec{k}|} = O\left(\frac{m}{k_0}\right) \tag{H.27}$$

and from (H.26), (H.27) thus immediately follows (H.25).

The relation (H.25) shows that the individual components of the four-vector of longitudinal polarization grow unboundedly in the high-energy limit since they behave like components of the corresponding four-momentum; let us emphasize, however, that the normalization $\varepsilon_L(k).\varepsilon_L^*(k) = -1$ still holds for an arbitrary k as it is defined by means of the indefinite Minkowski-space metric.

We will exhibit one more relation for polarization vectors of a massive vector particle which is frequently used in practical calculations, namely

$$\sum_{\lambda=1}^{3} \varepsilon_\mu(k, \lambda)\varepsilon_\nu^*(k, \lambda) = -g_{\mu\nu} + \frac{1}{m^2}k_\mu k_\nu \qquad (H.28)$$

(Notice that (H.28) is in a sense an analogy of the formulae (B.5), (B.6) valid for a Dirac particle). A proof of (H.28) is most easily performed in the following way. Since the $\varepsilon(k, \lambda)$ are four-vectors, the sum over polarizations on the left-hand side of eq. (H.28) must be a 2nd rank tensor depending on a single four-vector k. Denoting the considered polarization sum as $P_{\mu\nu}(k)$ one may therefore write

$$P_{\mu\nu}(k) = Ag_{\mu\nu} + Bk_\mu k_\nu \qquad (H.29)$$

where A, B are constants (because $k^2 = m^2$). Now it is sufficient to use a concrete form of the polarization vectors (which should be as simple as possible) for a conveniently chosen four-momentum k, e.g. for $k = (k_0, 0, 0, |\vec{k}|)$ (then one may employ e.g. the first two expressions from (H.11) and the corresponding particular value of (H.16)). With such a choice we obtain $P_{11}(k) = 1$, $P_{03}(k) = -k_0|\vec{k}|m^{-2}$ and using this in (H.29) we get immediately $A = -1$, $B = m^{-2}$ and eq. (H.28) is thus proved.

So far we have considered the Proca equations (H.1), (H.2) or (H.3) as equations for the wave function of a relativistic massive spin-1 particle. These equations may of course also be employed for the description of a corresponding classical free field. For simplicity we shall first consider the case of a real field (which corresponds to neutral particles upon quantization). The equations of motion (H.3) may be derived in a standard manner as the Euler-Lagrange equations corresponding to the Lagrangian density

$$\mathcal{L} = -\frac{1}{4}F_{\mu\nu}F^{\mu\nu} + \frac{1}{2}m^2 B_\mu B^\mu \qquad (H.30)$$

where $F_{\mu\nu} = \partial_\mu B_\nu - \partial_\nu B_\mu$. The classical field described by the Lagrangian (H.30) can be quantized canonically; in doing this, one has to keep in mind that the relevant independent dynamic variables are the space components B_j, $j = 1, 2, 3$. Details of the procedure of canonical quantization of the Proca field can be found e.g. in [21], [36], [38]. For the quantized field B_μ one may write an expansion into the the plane waves (H.6)

$$B_\mu(x) = \int \frac{d^3k}{(2\pi)^{3/2}(2k_0)^{1/2}} \sum_{\lambda=1}^{3} [a(k, \lambda)\varepsilon_\mu(k, \lambda)e^{-ikx} + a^+(k, \lambda)\varepsilon_\mu^*(k, \lambda)e^{ikx}] \qquad (H.31)$$

where the polarization vectors $\varepsilon(k, \lambda)$ satisfy the conditions (H.8), (H.13) and the normalization factor $N(k)$ in (H.6) is chosen so that the canonical commutation relations for $B_j(x)$ and the corresponding conjugate momenta imply the following commutation relations for the annihilation and creation operators in the decomposition (H.31):

$$[a(k, \lambda), \ a^+(k', \lambda')] = \delta_{\lambda\lambda'}\delta^3(\vec{k} - \vec{k}') \tag{H.32}$$

Now we are going to calculate the corresponding Feynman propagator. One may start with its usual representation in terms of time-ordered product of a pair of field operators, i.e. define

$$i\mathcal{D}_{\mu\nu}(x - y) =< 0|T(B_\mu(x)B_\nu(y))|0 > \tag{H.33}$$

where

$$T(B_\mu(x)B_\nu(y)) = \vartheta(x_0 - y_0)B_\mu(x)B_\nu(y) + \vartheta(y_0 - x_0)B_\nu(y)B_\mu(x) \tag{H.34}$$

To compute the expression on the right-hand side of (H.33) one employs the decomposition (H.31), commutators of the type (H.32) and the formula (H.28). Standard manipulations then lead to a result for the propagator $\mathcal{D}_{\mu\nu}(x - y)$ which contains non-covariant terms proportional to $g_{0\mu}g_{0\nu}\delta^4(x - y)$ (see e.g. [21], §3.2.3, §5.1.7 and [38]). In general, one may expect such contact terms to be present in the propagator, because the time-ordering operation T in (H.33) is not, *a priori*, strictly defined for $x_0 = y_0$; the ϑ-function in the conventional definition (H.34) has unique meaning as a generalized function but relativistic covariance of (H.34) is not manifest. Let us however remark that the above-mentioned problem does not occur in the massless case (i.e. for the electromagnetic field). It is clear that in view of the above-mentioned ambiguity of the massive vector-boson propagator for $x = y$ one has to postulate an additional requirement of relativistic covariance (it is usually formulated as replacing the symbol T in (H.33) by an appropriate covariant time-ordering T^* see e.g. [21]). On the other hand, the Feynman propagator of a massive vector field may also be viewed as the causal Green function of the Proca equation (H.3); a practical computation of the covariant propagator function $\mathcal{D}_{\mu\nu}(x)$ is performed most easily by utilizing this connection. Thus, one has to solve the equation

$$(\Box + m^2)\mathcal{D}_\nu^\mu(x) - \partial^\mu(\partial_\lambda \mathcal{D}_\nu^\lambda(x)) = g_\nu^\mu\delta^4(x) \tag{H.35}$$

(a solution of Eq. (H.35), if it exists, is automatically a 2nd rank tensor with respect to Lorentz transformations). Performing Fourier transformation in Eq. (H.35), i.e. introducing the function $D_{\mu\nu}(k)$ defined by

$$\mathcal{D}_{\mu\nu}(x) = \int \frac{d^4k}{(2\pi)^4} e^{ikx} D_{\mu\nu}(k) \tag{H.36}$$

one gets from Eq. (H.35) the system of algebraic equations

$$(-k^2 + m^2)D_\nu^\mu(k) + k^\mu k_\lambda D_\nu^\lambda(k) = g_\nu^\mu \tag{H.37}$$

or

$$L^\mu_\lambda D^\lambda_\nu = g^\mu_\nu \tag{H.38}$$

where

$$L^\mu_\lambda = (-k^2 + m^2)g^\mu_\lambda + k^\mu k_\lambda \tag{H.39}$$

The $D^{\rho\sigma}(k)$ is a 2nd rank tensor (depending on a single four-vector k) and thus it may in general be written as

$$D^{\rho\sigma}(k) = D_T(k^2)P_T^{\rho\sigma}(k) + D_L(k^2)P_L^{\rho\sigma}(k) \tag{H.40}$$

where

$$P_T^{\rho\sigma} = g^{\rho\sigma} - \frac{k^\rho k^\sigma}{k^2}$$
$$P_L^{\rho\sigma} = \frac{k^\rho k^\sigma}{k^2} \tag{H.41}$$

Denoting as P_T and P_L the matrices with elements given by the mixed components of the tensors (H.41), it is easy to find that

$$P_T^2 = P_T, \quad P_L^2 = P_L, \quad P_T P_L = P_L P_T = 0, \tag{H.42}$$

i.e. the matrices P_T and P_L are orthogonal projectors - this is a substantial advantage of the parametrization (H.40). The matrix L defined by (H.39) may be decomposed in an analogous way:

$$L = (-k^2 + m^2)P_T + m^2 P_L \tag{H.43}$$

Employing the relations (H.42) together with (H.40) and (H.43) it is now easy to solve the matrix equation (H.38); since the unit matrix in its right-hand side may be written as $P_T + P_L$ one readily gets

$$D_T = \frac{1}{-k^2 + m^2}, \quad D_L = \frac{1}{m^2} \tag{H.44}$$

(for $k^2 \neq m^2$). The ambiguity corresponding to a potential singularity at $k^2 = m^2$ is removed by defining the causal Green function in a standard way, i.e. by the familiar replacement $m^2 \to m^2 - i\varepsilon$. According to (H.40), (H.41) and (H.44) one thus gets the final result for the Feynman propagator of the massive vector field in momentum space:

$$D_{\mu\nu}(k) = \frac{-g_{\mu\nu} + m^{-2}k_\mu k_\nu}{k^2 - m^2 + i\varepsilon} \tag{H.45}$$

In closing this appendix let us also remark that for a classical *complex* vector field one has to write the corresponding free Lagrangian as

$$\mathcal{L} = -\frac{1}{2}(\partial_\mu B_\nu - \partial_\nu B_\mu)(\partial^\mu B^{\nu*} - \partial^\nu B^{\mu*}) + m^2 B_\mu B^{\mu*} \tag{H.46}$$

or for a quantized non-hermitean field (i.e. a field corresponding to charged particles), in the form

$$\mathcal{L} = -\frac{1}{2}(\partial_\mu B_\nu^- - \partial_\nu B_\mu^-)(\partial^\mu B^{+\nu} - \partial^\nu B^{+\mu}) + m^2 B_\mu^- B^{+\mu} \qquad \text{(H.47)}$$

where the B_μ^- and B_μ^+ are related by means of hermitean conjugation. The change of coefficients in (H.46) in comparison with (H.30) is of course due to the fact that for a complex field, the B_j and B_j^* are independent dynamical variables. For a charged vector field one also has to modify plane-wave decompositions of the type (H.31) (cf. e.g. the expressions (B.4) for a Dirac field). The formula (H.45) for the Feynman propagator (which in the case of charged vector bosons is defined by means of time-ordered product of the fields $B_\mu^-(x)$ and $B_\nu^+(y)$) is not changed. Thus, in practical calculations of Feynman diagrams involving charged intermediate vector bosons of weak interactions, an internal IVB line labelled e.g. by W^- corresponds to the same propagator as that labelled by W^+ and the relevant expression is always given by (H.45).

Appendix I

THE INTERACTION WWZ

We are going to prove first a basic statement on the direct interaction of three vector bosons $W^\pm$, Z set forth in Section 5.2, namely:

Leading divergences arising in the limit $E \to \infty$ in tree-level diagrams (of binary processes) involving interaction vertices WWZ vanish for an arbitrary combination of polarizations of the external $W^\pm$ and Z if and only if the interaction WWZ is of the Yang-Mills type, i.e. the vertex in Fig. 15 is given by the expression (see (5.14), (4.15))

$$\mathcal{V}^{(YM)}_{\lambda\mu\nu}(k,p,q) = g_{WWZ} V^{(YM)}_{\lambda\mu\nu}(k,p,q) \tag{I.1}$$

where

$$V^{(YM)}_{\lambda\mu\nu}(k,p,q) = (p-q)_\lambda g_{\mu\nu} + (q-k)_\mu g_{\lambda\nu} + (k-p)_\nu g_{\lambda\mu} \tag{I.2}$$

and g_{WWZ} is a (real) coupling constant.

Further, at the end of this appendix we will show how one can generalize the corresponding statement concerning the electromagnetic interaction $WW\gamma$ of the Yang-Mills type which we have derived in Chapter 4.

A proof of the first part of the above assertion (stating that the Yang-Mills structure (5.2) is a *sufficient* condition for an elimination of the corresponding divergences) is based on applications of the 't Hooft identity (4.19). Since we have already used such a technique in several particular examples in the main text, we leave a formulation of a proof of the first part of our statement to the reader.

Now we are going to prove the more difficult part of the statement, namely that the Yang-Mills structure (I.2) of the WWZ interaction is a *necessary* condition for an elimination of the leading high-energy divergences in the corresponding tree graphs. Of course, in doing this we will only consider interaction terms satisfying the constraint (5.5), i.e.

$$\dim \mathcal{L}_{WWZ} \leq 4 \tag{I.3}$$

It is obvious that a Lorentz-invariant interaction of three vector bosons fulfilling the condition (I.3) must involve just one derivative of a vector-boson field (the corresponding coupling constant is then of course dimensionless). In momentum space, this means that the interaction vertex shown in Fig. 15 represents a linear polynomial with respect to the four-momenta k, p, q. In fact only two of these four-momenta are independent since $k + p + q = 0$. Choosing e.g. the k and p to be independent variables, the most general linear polynomial representing the interaction vertex WWZ may be written as

$$
\begin{aligned}
\mathcal{V}_{\lambda\mu\nu}(k, p, q) &= \\
&= (Ak_\lambda + Bp_\lambda)g_{\mu\nu} + (Ck_\mu + Dp_\mu)g_{\lambda\nu} + (Ek_\nu + Fp_\nu)g_{\lambda\mu} \\
&+ G\varepsilon_{\lambda\mu\nu\rho}k^\rho + H\varepsilon_{\lambda\mu\nu\rho}p^\rho
\end{aligned}
\tag{I.4}
$$

For comparison, the expression for the Yang-Mills vertex (I.2) may be written (using the four-momentum conservation $q = -(k + p)$) as

$$
\mathcal{V}^{(YM)}_{\lambda\mu\nu}(k, p, q) = (k + 2p)_\lambda g_{\mu\nu} + (-2k - p)_\mu g_{\lambda\nu} + (k - p)_\nu g_{\lambda\mu}
\tag{I.5}
$$

On the general interaction vertex (I.4) one may now impose constraints following from the requirement of a suppression of the leading high-energy divergences in relevant tree-level Feynman diagrams. For this purpose we will consider 3 different configurations of the vector boson lines $W^\pm$, Z, such that the Z, W^+ or W^- label consecutively an internal line outgoing from the WWZ vertex in a Feynman graph (with the other two vector bosons corresponding to external lines). These 3 configurations correspond e.g. to processes $e^+e^- \to W^+W^-$, $\bar{\nu}e^- \to W^-Z$ and $\nu e^+ \to W^+Z$ (see Fig. 39).

a) First we shall examine leading power-like divergences arising in the limit $E \to \infty$ from the diagram in Fig. 39(a). Obviously, the worst divergence comes in any case (i.e. for an arbitrary combination of the $W^\pm$ polarizations) from the longitudinal part of the Z propagator which is proportional to $m_Z^{-2}q^\mu q^\nu$ Acting with the q^α on the leptonic vertex, the electron mass m is factorized, which compensates one factor of m_Z^{-1}; however, there remains another *a priori* uncompensated factor m_Z^{-1} which may cause the degree of divergence of the diagram (a) for $E \to \infty$ in general to be higher than that of any other tree graph contributing to $e^+e^- \to W^+W^-$. (Such an argument may be used in all the considered cases, i.e. for the diagrams (b) and (c) as well, and we will keep it in mind implicitly in what follows in estimating high-energy behaviour of the leading divergent terms.)

Thus, for an arbitrary combination of polarizations of the final-state W's the leading divergence in question resides in the expression

$$
m_Z^{-1}q^\nu \mathcal{V}_{\lambda\mu\nu}(k, p, q)\varepsilon^{*\mu}(p)\varepsilon^{*\lambda}(k)
\tag{I.6}
$$

Substituting the general parametrization (I.4) into (I.6) and using the conservation law $q = -(k + p)$, then for the leading term contained in (I.6) after a simple manipulation one gets

$$
m_Z^{-1}\Big[\; - (B + C)(k.\varepsilon^*(p))(p.\varepsilon^*(k)) \\
- (E + F)(k.p)(\varepsilon^*(k).\varepsilon^*(p)) \\
+ (G - H)\varepsilon_{\lambda\mu\nu\rho}k^\nu p^\rho \varepsilon^{*\lambda}(k)\varepsilon^{*\mu}(p) \Big]
\tag{I.7}
$$

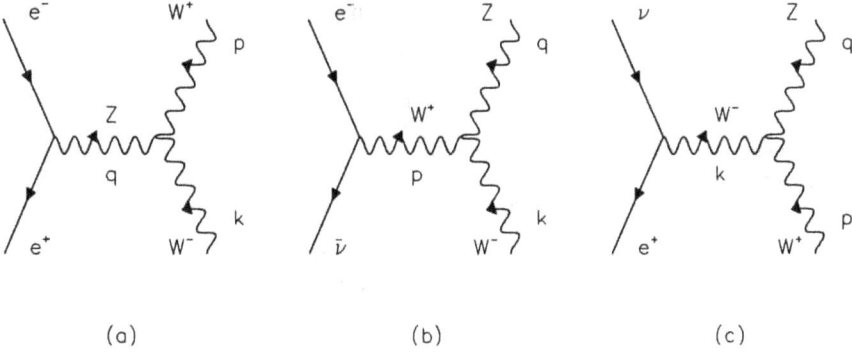

Fig. 39. Tree diagrams of processes (a) $e^- e^+ \to W^- W^+$ (b) $\bar{\nu} e^- \to W^- Z$ (c) $\nu e^+ \to W^+ Z$ involving the interaction vertex WWZ.

(In deriving (I.7) we have of course also utilized the relations $k^2 = m_W^2$, $p^2 = m_W^2$, $k.\varepsilon^*(k) = 0$, $p.\varepsilon^*(p) = 0$ and we have neglected non-leading terms in which m_W^2 is factorized.) The requirement of an elimination of leading divergences in the diagram (a) thus means that the coefficients of all the independent kinematical structures in (I.7) should vanish. So we get the conditions

$$B + C = 0 \tag{I.8}$$

$$E + F = 0 \tag{I.9}$$

$$G - H = 0 \tag{I.10}$$

b) We will now examine the diagram in Fig. 39(b). Here, a potential leading divergence comes from the expression

$$m_W^{-1} p^\mu \mathcal{V}_{\lambda\mu\nu}(k, p, q) \varepsilon^{*\lambda}(k) \varepsilon^{*\nu}(q) \tag{I.11}$$

Substituting (I.4) into (I.11) and using $p = -(k + q)$, then similarly to the preceding case, for the leading term contained in (I.11) we obtain

$$m_W^{-1} \left[\begin{array}{l} (B - E + F)(k.\varepsilon^*(q))(q.\varepsilon^*(k)) \\ + \ (-C + 2D)(k.q)(\varepsilon^*(k).\varepsilon^*(q)) \\ + \ G\varepsilon_{\lambda\mu\nu\rho}k^\mu q^\rho \varepsilon^{*\lambda}(k)\varepsilon^{*\nu}(q) \end{array} \right] \tag{I.12}$$

The requirement of an elimination of leading divergences in the diagram (b) thus yields the conditions

$$B - E + F = 0 \tag{I.13}$$

$$-C + 2D = 0 \tag{I.14}$$

$$G = 0 \tag{I.15}$$

c) Finally, for the diagram in Fig. 39(c) the corresponding leading divergence comes from

$$m_W^{-1} k^\lambda V_{\lambda\mu\nu}(k, p, q) \varepsilon^{*\mu}(p) \varepsilon^{*\nu}(q) \tag{I.16}$$

Substituting (I.4) into (I.16) and using $k = -(p + q)$, then in an analogous manner as in the preceding cases for the leading term involved in (I.16) we get

$$
m_W^{-1} \left[\begin{aligned}
& (C + E - F)(p.\varepsilon^*(q))(q.\varepsilon^*(p)) \\
+\ & (2A - B)(p.q)(\varepsilon^*(p).\varepsilon^*(q)) \\
+\ & H\varepsilon_{\lambda\mu\nu\rho} p^\lambda q^\rho \varepsilon^{*\mu}(p)\varepsilon^{*\nu}(q)
\end{aligned} \right] \tag{I.17}
$$

The requirement of an elimination of leading divergences in the diagram (c) thus yields the conditions

$$C + E - F = 0 \tag{I.18}$$

$$2A - B = 0 \tag{I.19}$$

$$H = 0 \tag{I.20}$$

Thus, in the first place we see that two of the eight unknown parameters in (I.4) must vanish if one wants to suppress all leading high-energy divergences in the diagrams in Fig. 39, namely (see (I.10), (I.15), (I.20))

$$G = H = 0 \tag{I.21}$$

In other words, the two terms in the expression for the WWZ interaction vertex involving the Levi-Civita tensor $\varepsilon_{\lambda\mu\nu\rho}$ are identically zero. For the remaining six unknowns $A, \ldots, F$ we have obtained a system of six conditions (I.8), (I.9), (I.13), (I.14), (I.18) and (I.19). For convenience, let us summarize these equations here:

$$
\begin{aligned}
B + C &= 0 \\
E + F &= 0 \\
B - E + F &= 0 \\
-C + 2D &= 0 \\
C + E - F &= 0 \\
2A - B &= 0
\end{aligned} \tag{I.22}
$$

It is easy to find that the solution of the system (I.22) is unique, up to a one-parametric freedom in choosing arbitrarily one of the unknowns (e.g. A), namely

$$B = 2A, \quad C = -2A, \quad D = -A, \quad E = A, \quad F = -A \qquad (I.23)$$

The result (I.23) means that the most general expression (I.4) constrained to satisfy our conditions has the form

$$\mathcal{V}_{\lambda\mu\nu}(k,p,q) = A\Big[(k + 2p)_\lambda g_{\mu\nu} + (-2k - p)_\mu g_{\lambda\nu} + (k - 2p)_\nu g_{\lambda\mu}\Big] \qquad (I.24)$$

with A being an arbitrary constant. This, however, is the interaction vertex of the type (I.5) and $A = g_{WWZ}$ in the notation of (I.1). Thus we see that the necessary condition for eliminating the leading high-energy divergences in the particular graphs in Fig. 39 is that the WWZ interaction be of the Yang-Mills type; thus it is a necessary condition for suppressing unwanted divergences in a general case. Our statement is thereby proved.

The following comment on the obtained results is in order: In the general expression (I.4) we have started from, it has not been necessary to assume *a priori* that the parameters $A,\ldots,H$ are real; however, from (I.24) it is clear that the parameter A must be real (and, according to (I.23), the same is then true for the rest) for the corresponding interaction Lagrangian to be hermitean (cf. (5.13), (5.14)). Thus, according to (I.21) and (I.23), the solution of the considered problem admits only real values of the parameters in (I.4).

To close this appendix, we will make an important comment concerning the electromagnetic interaction $WW\gamma$. In Chapter 4 we derived the Yang-Mills structure of the corresponding interaction vertex, starting from the electromagnetic interaction (4.7) involving one free parameter κ (we then used *a priori* also some restrictions which follow from imposing the discrete symmetries C, P and T). The question naturally arises, as to whether it would be possible to derive the Yang-Mills interaction $WW\gamma$ in a manner analogous to that employed here in the WWZ case. The answer to this question is yes: The procedure described in this appendix may easily be generalized and used almost without any change for the electromagnetic interaction $WW\gamma$. To this end it is sufficient to consider a general parametrization of the type (I.4) and diagrams analogous to those in Fig. 39 with the Z lines being replaced by photons; one has to realize that for a diagram of the type (a) (involving an internal photon line) a corresponding argument has to be reformulated: In such a case one has to require an elimination of the longitudinal part of the photon propagator on the basis of electromagnetic gauge independence (mind that the longitudinal part of photon propagator may depend on a gauge-fixing parameter) and not because of suppressing an offending high-energy divergence. (Strictly speaking, for the considered graph the required effect occurs automatically owing to the current conservation in the leptonic vertex; thus, in order to draw indeed a non-trivial constraint for $WW\gamma$ from electromagnetic gauge-independence, one should instead consider e.g. tree diagrams involving two $WW\gamma$ vertices - an obvious example is provided by elastic WW

scattering.) Thus, although a physical origin of the relevant condition formulated for an electromagnetic diagram of the type (a) (in a broader sense) is different from the case of the WWZ interaction, it is clear that *technically* such a condition leads to the same equations for parameters in an expression of the type (I.4), i.e. we thus recover the relations (I.8), (I.9) and (I.10). For photonic diagrams of the type (b) or (c) the corresponding conditions are formulated in the same way as in the case of WWZ interaction (i.e. by requiring a suppression of the would-be leading high-energy divergences). The above remark concerning the $WW\gamma$ interaction thus provides an interesting non-trivial generalization of the arguments used in Chapter 4.

Appendix J

HIGH-ENERGY BEHAVIOUR
OF SOME TREE DIAGRAMS

In this appendix we summarize formulae for the leading and next-to-leading asymptotic terms corresponding to the limit $E \to \infty$ in contributions of some important tree-level Feynman diagrams discussed in the main text. In more complicated cases we give a brief derivation as well.

1. *The process $e^+ e^- \to W_L^+ W_L^-$*

   (a) The contribution of the diagram in Fig. 17(a) may be written as

   $$\mathcal{M}_{17a} = -\frac{g^2}{4m_W^2}\bar{v}(l)\not{p}(1-\gamma_5)u(k)$$
   $$- \frac{g^2}{4m_W^2}m\bar{v}(l)(1-\gamma_5)u(k) + O(1) \tag{J.1}$$

   A derivation of this result is left to the reader as an easy instructive exercise (see the problem 3.6 in Chapter 3).

   (b) The contribution of Fig. 17(b) contains only a quadratically diverging term (see (4.34) or (5.22)).

   (c) The contribution of Fig. 17(c) will now be worked out in more detail. The starting point of our calculation is the expression

   $$i\mathcal{M}_{17c} = i^3 g_{WWZ}\bar{v}(l)\left(g_L\gamma_\rho\frac{1-\gamma_5}{2} + g_R\gamma_\rho\frac{1+\gamma_5}{2}\right)u(k) \times$$
   $$\times \frac{-g^{\rho\nu} + m_Z^{-2}q^\rho q^\nu}{q^2 - m_Z^2}V_{\lambda\mu\nu}(p,r,q)\varepsilon_L^\lambda(p)\varepsilon_L^\mu(r) \tag{J.2}$$

   (here and in what follows we take into account that the vector of longitudinal polarization is real - see (H.16)). Employing the cyclic property of the $V_{\lambda\mu\nu}(p,r,q)$ (see Eq. (4.18)) and 't Hooft identity (4.19) it is easy to show that the longitudinal part of the Z propagator does not contribute (this is even true for an arbitrary combination of $W^{\pm}$ polarizations). Further using the standard decomposition (H.25), (J.2) may be rewritten as

$$
\begin{aligned}
\mathcal{M}_{17c} &= g_{WWZ}\frac{1}{m_W^2}\bar{v}(l)\left(g_L\gamma^\nu\frac{1-\gamma_5}{2}+g_R\gamma^\nu\frac{1+\gamma_5}{2}\right)u(k)\times \\
&\times \frac{1}{s-m_Z^2}V_{\lambda\mu\nu}(p,r,q)p^\lambda r^\mu + O(1)
\end{aligned} \tag{J.3}
$$

where $s = q^2 = (k+l)^2$. Using again the 't Hooft identity in (J.3), we get, after a short manipulation

$$
\begin{aligned}
\mathcal{M}_{17c} &= -g_{WWZ}\frac{1}{m_W^2}\bar{v}(l)\left(g_L\gamma^\nu\frac{1-\gamma_5}{2}+g_R\gamma^\nu\frac{1+\gamma_5}{2}\right)u(k)\times \\
&\times (p_\nu + \tfrac{1}{2}q_\nu) + O(1)
\end{aligned} \tag{J.4}
$$

An application of the Dirac equation in (J.4) finally leads to the result

$$
\begin{aligned}
\mathcal{M}_{17c} = &-\frac{1}{2m_W^2}g_{WWZ}g_L\bar{v}(l)\not{p}(1-\gamma_5)u(k) \\
&-\frac{1}{2m_W^2}g_{WWZ}g_R\bar{v}(l)\not{p}(1+\gamma_5)u(k) \\
&+\frac{m}{2m_W^2}g_{WWZ}(g_L-g_R)\bar{v}(l)\gamma_5 u(k) + O(1)
\end{aligned} \tag{J.5}
$$

2. *The process $\bar{v}e \to W_L^- Z_L$*

(a) The contribution of Fig. 18(a) is given by

$$
\begin{aligned}
\mathcal{M}_{18a} = &-\frac{g}{2\sqrt{2}}g_L\frac{1}{m_W m_Z}\bar{v}(l)\not{p}(1-\gamma_5)u(k) \\
&+\frac{g}{2\sqrt{2}}g_R\frac{m}{m_W m_Z}\bar{v}(l)(1+\gamma_5)u(k) + O(1)
\end{aligned} \tag{J.6}
$$

(b) For the diagram in Fig. 18(b) one has

$$
\begin{aligned}
\mathcal{M}_{18b} = &\frac{g}{2\sqrt{2}}g_{\nu\nu Z}\frac{1}{m_W m_Z}\bar{v}(l)\not{p}(1-\gamma_5)u(k) \\
&-\frac{g}{2\sqrt{2}}g_{\nu\nu Z}\frac{m}{m_W m_Z}\bar{v}(l)(1+\gamma_5)u(k) + O(1)
\end{aligned} \tag{J.7}
$$

A derivation of the formulae (J.6) and (J.7) is straightforward and we leave it to the reader as an instructive exercise.

(c) The evaluation of Fig. 18(c) is slightly more complicated and we will therefore indicate here at least its most important steps. As a starting point, let us take the basic expression

$$i\mathcal{M}_{18c} = i^3 \frac{g}{2\sqrt{2}} g_{WWZ} \bar{v}(l)\gamma_\rho(1 - \gamma_5)u(k) \times$$

$$\times \frac{-g^{\rho\nu} + m_W^{-2}q^\rho q^\nu}{q^2 - m_W^2} V_{\lambda\mu\nu}(p,r,q)\varepsilon_L^\lambda(p)\varepsilon_L^\mu(r) , \qquad (J.8)$$

that is

$$\mathcal{M}_{18c} = \mathcal{M}_{18c}^{(1)} + \mathcal{M}_{18c}^{(2)} , \qquad (J.9)$$

where the $\mathcal{M}_{18c}^{(1)}$ and $\mathcal{M}_{18c}^{(2)}$ respectively correspond to the diagonal and longitudinal parts of the W propagator in (J.8). First we will compute the $\mathcal{M}_{18c}^{(2)}$. Employing Eq. (4.18) and (4.19) one may easily show that

$$q^\nu V_{\lambda\mu\nu}(p,r,q)\varepsilon_L^\lambda(p)\varepsilon_L^\mu(r) = (m_W^2 - m_Z^2)\varepsilon_L(p).\varepsilon_L(r) \qquad (J.10)$$

Also using the Dirac equation and the decomposition (H.25), after a short manipulation from (J.8) and (J.10) we get

$$\mathcal{M}_{18c}^{(2)} = \frac{g g_{WWZ}}{4\sqrt{2}} \frac{m}{m_W m_Z}\left(1 - \frac{m_Z^2}{m_W^2}\right)\bar{v}(l)(1 + \gamma_5)u(k) + O(1) \qquad (J.11)$$

From the last expression it is clear that the $\mathcal{M}_{18c}^{(2)}$ contains terms at most linearly divergent for $E \to \infty$. The calculation of the part $\mathcal{M}_{18c}^{(1)}$ is analogous to the case of Fig. 17(c). Again we use the decomposition (H.25), the 't Hooft identity (4.19) and the Dirac equation and after simple algebraic manipulations we obtain the result

$$\mathcal{M}_{18c}^{(1)} = -\frac{g g_{WWZ}}{2\sqrt{2}} \frac{1}{m_W m_Z}\bar{v}(l)\not{p}(1 - \gamma_5)u(k)$$

$$+ \frac{g g_{WWZ}}{4\sqrt{2}} \frac{m}{m_W m_Z}\bar{v}(l)(1 + \gamma_5)u(k)$$

$$+ O(1) \qquad (J.12)$$

According to (J.9), (J.11) and (J.12) we thus have for the whole contribution of Fig. 18(c)

$$\mathcal{M}_{18c} = -\frac{g g_{WWZ}}{2\sqrt{2}} \frac{1}{m_W m_Z}\bar{v}(l)\not{p}(1 - \gamma_5)u(k)$$

$$+ \frac{g g_{WWZ}}{2\sqrt{2}} \frac{m}{m_W m_Z}\left(1 - \frac{m_Z^2}{2m_W^2}\right)\bar{v}(l)(1 + \gamma_5)u(k)$$

$$+ O(1) \qquad (J.13)$$

3. *The process* $W_L^- W_L^- \to W_L^- W_L^-$

Here we shall examine contributions of the diagrams in Fig. 7 (photon exchange), Fig. 19 (Z exchange) and Fig. 20(c) (direct interaction of four vector bosons) and prove the relation (5.53). All these diagrams are summarized in Fig. 20. (There are of course also contributions of neutral scalar boson exchange (see Fig. 25); the corresponding calculation is relatively simple and we leave it to the reader as an exercise - see the problem 5.7).

(a) Let us first consider the contribution of Fig. 7(a). The corresponding amplitude is given by

$$
\begin{aligned}
i\mathcal{M}_{7a} &= i^3 e^2 V_{\lambda\mu\rho}(p, -k, q)\frac{-g^{\rho\sigma}}{q^2}V_{\sigma\tau\omega}(-q, r, -l) \times \\
&\quad \times \; \varepsilon_L^\lambda(p)\varepsilon_L^\mu(k)\varepsilon_L^\tau(r)\varepsilon_L^\omega(l)
\end{aligned}
\tag{J.14}
$$

The photon propagator in (J.14) corresponds to the Feynman gauge and for the $WW\gamma$ vertices we have used the rule that an incoming W^- is equivalent to an outgoing W^+ with opposite four-momentum (see Chapter 4, the remark following Eq. (4.14)); in each case one has to maintain an order of the momentum variables in the function $\dot V$ and of the corresponding indices ($\gamma W^- W^+$ or an arbitrary cyclic permutation). For the vectors of longitudinal polarizations according to (H.25) one may write

$$
\varepsilon_L^\lambda(p) = \frac{1}{m_W}p^\lambda + \Delta^\lambda(p)
\tag{J.15}
$$

etc. where the remainder $\Delta^\lambda(p)$ is of the order $O(m_W/E)$. Since $p.\varepsilon_L(p) = 0$ and $p^2 = m_W^2$, a useful identity follows immediately from (J.15), namely

$$
p.\Delta(p) = -m_W
\tag{J.16}
$$

Our goal now is to isolate leading and next-to-leading asymptotic terms in Eq. (J.14), i.e. the terms of the order $O(E^4/m_W^4)$ and $O(E^2/m_W^2)$ for $E \to \infty$. Substituting into (J.14) a decomposition of the type (J.15) for each polarization vector, the amplitude $\mathcal{M}_{7a}$ becomes a sum of 16 terms; the first of them contains the product

$$
m_W^{-4} p^\lambda k^\mu r^\tau l^\omega \; ,
\tag{J.17}
$$

the next one is proportional to

$$
m_W^{-3} \Delta^\lambda(p) k^\mu r^\tau l^\omega \; ,
\tag{J.18}
$$

etc. It is obvious that the leading (i.e. quartic) divergence may only come from the term involving (J.17) (this of course contains a part of quadratic divergences as well). Further quadratic divergences arise in terms involving products of the type (J.18) (there are four such terms). All the other contributions to $\mathcal{M}_{7a}$ are already of the order $O(1)$ for $E \to \infty$, as one may easily guess on the basis of the asymptotic behaviour of the leading term and of the remainder in the decomposition (J.15). Following these simple considerations we thus get from (J.14)

$$\mathcal{M}_{7a} = \frac{e^2}{t} \sum_{j=1}^{5} X_j + O(1) \tag{J.19}$$

where

$$X_1 = \frac{1}{4m_W^4}(t^3 + 2st^2) - \frac{1}{m_W^2}t^2$$

$$X_2 = \frac{1}{4m_W^2}(t^2 - 2st) + \frac{t^2}{2m_W^3}p.\Delta(k) + \frac{t^2}{m_W^3}r.\Delta(k)$$

$$X_3 = \frac{1}{4m_W^2}(3t^2 + 2tu) + \frac{t^2}{m_W^3}l.\Delta(p) + \frac{t^2}{2m_W^3}k.\Delta(p)$$

$$X_4 = \frac{1}{4m_W^2}(t^2 - 2st) + \frac{t^2}{m_W^3}p.\Delta(l) + \frac{t^2}{2m_W^3}r.\Delta(l)$$

$$X_5 = \frac{1}{4m_W^2}(3t^2 + 2tu) + \frac{t^2}{m_W^3}k.\Delta(r) + \frac{t^2}{2m_W^3}l.\Delta(r)$$

$$\tag{J.20}$$

In (J.19) and (J.20) we have used the standard notation (see Fig. 7)

$$s = (k+l)^2 = (p+r)^2$$
$$t = (k-p)^2 = (l-r)^2 = q^2$$
$$u = (k-r)^2 = (l-p)^2 = Q^2$$

For a derivation of the expression for the X_1 it is sufficient to use the 't Hooft identity (4.19) (and of course take into account that $k^2 = l^2 = p^2 = r^2 = m_W^2$). To derive the expressions for $X_2, ..., X_5$ in (J.20) one has to use, in addition, identities of the type (J.16) for the corresponding four-momenta; the rest is a straightforward algebra.

The diagram in Fig. 7(b) corresponds to an interchange $p \leftrightarrow r$, i.e. also $t \leftrightarrow u$. Performing this (and also using $s + t + u = 4m_W^2$) for the whole contribution of Fig. 7 we get

$$\frac{1}{e^2}(\mathcal{M}_{7a} + \mathcal{M}_{7b}) = \frac{1}{4m_W^4}(t^2 + u^2 - 2s^2) - \frac{2s}{m_W^2}$$

$$+ \frac{1}{2m_W^3}(t + 2u)(k.\Delta(p) + p.\Delta(k) + l.\Delta(r) + r.\Delta(l))$$

$$+ \ \frac{1}{2m_W^3}(u + 2t)(k.\Delta(r) + r.\Delta(k) + l.\Delta(p) + p.\Delta(l))$$

$$+ \ O(1) \tag{J.21}$$

(b) We shall now examine the contribution of the diagrams in Fig. 19(a), (b) which correspond to the Z exchange. Let us first consider the diagram (a). The WWZ interaction is of the Yang-Mills type (i.e. it has the same structure as the $WW\gamma$ vertex - see Eq. (5.13), (5.14) and (5.69)) and the corresponding amplitude $\mathcal{M}_{19a}$ is obtained from the $\mathcal{M}_{7a}$ by replacing e^2 with g_{WWZ}^2 and using the Z propagator instead of photon propagator. It is easy to show that the longitudinal part of the Z propagator does not contribute, as in Fig. 17(c) (see the remark following Eq. (J.2)). Thus only the diagonal part of the Z propagator contributes to the amplitude $\mathcal{M}_{19a}$; it means that an evaluation of the $\mathcal{M}_{19a}$ is essentially identical with the case of the $\mathcal{M}_{7a}$ - one only has to replace the t^{-1} in (J.19) by $(t - m_Z^2)^{-1}$ (and of course also replace e^2 by g_{WWZ}^2). Thus we have

$$\mathcal{M}_{19a} = g_{WWZ}^2 \frac{1}{t - m_Z^2} \sum_{j=1}^{5} X_j + O(1) , \tag{J.22}$$

where the X_j, $j = 1, ..., 5$ are given by the expressions (J.20). The contribution of Fig. 19(b) is then obtained from (J.22) by interchanging $p \leftrightarrow r$. After a simple algebraic manipulation we thus get finally

$$\frac{1}{g_{WWZ}^2}(\mathcal{M}_{19a} + \mathcal{M}_{19b}) =$$

$$= \ \frac{1}{4m_W^4}(t^2 + u^2 - 2s^2) - \frac{2s}{m_W^2} + \frac{3}{4}\frac{m_Z^2}{m_W^4}s$$

$$+ \ \frac{1}{2m_W^3}(t + 2u)(k.\Delta(p) + p.\Delta(k) + l.\Delta(r) + r.\Delta(l))$$

$$+ \ \frac{1}{2m_W^3}(u + 2t)(k.\Delta(r) + r.\Delta(k) + l.\Delta(p) + p.\Delta(l))$$

$$+ \ O(1) \tag{J.23}$$

Notice that Eq. (J.23) contains, in comparison with (J.21), some extra quadratic divergences (see the term proportional to m_Z^2 in (J.23)). This of course is a consequence of replacing t^{-1} by $(t - m_Z^2)^{-1}$ when passing from (J.19) to (J.22); these extra quadratic terms arise from the original quartic terms in Eq. (J.21) upon such a replacement.

(c) Finally, we shall examine the contribution of Fig. 20(c). Let us consider a general interaction of the type $WWWW$ parametrized by coupling constants a, b (see (5.49)). Using the decomposition (J.15) we then get (proceeding in an analogous way as before) the result

$$
\begin{aligned}
\mathcal{M}_{20c} \;=\; & 2a\Big[\frac{1}{4m_W^4}(t^2+u^2)+\frac{s}{m_W^2} \\
& -\frac{t}{2m_W^3}(k.\Delta(p)+p.\Delta(k)+l.\Delta(r)+r.\Delta(l)) \\
& -\frac{u}{2m_W^3}(k.\Delta(r)+r.\Delta(k)+l.\Delta(p)+p.\Delta(l))\Big] \\
& +4b\Big[\frac{1}{4m_W^4}s^2+\frac{s}{m_W^2} \\
& -\frac{1}{2m_W^3}(t+u)(k.\Delta(p)+p.\Delta(k)+l.\Delta(r)+r.\Delta(l) \\
& +k.\Delta(r)+r.\Delta(k)+l.\Delta(p)+p.\Delta(l))\Big] \\
& +O(1)
\end{aligned}
\tag{J.24}
$$

Substituting into Eq. (J.21) and (J.23) the "right" values of coupling constants

$$
e = g\sin\vartheta_W , \qquad g_{WWZ} = g\cos\vartheta_W
\tag{J.25}
$$

(see (5.36), (5.37)) the condition of a cancellation of the leading (quartic) divergences yields (cf. (5.51))

$$
a = -\frac{1}{2}g^2 , \qquad b = \frac{1}{2}g^2
\tag{J.26}
$$

Using the values (J.26) in the expression (J.24) we then get, after a simple manipulation

$$
\begin{aligned}
-\frac{1}{g^2}\mathcal{M}_{20c} \;=\; & \frac{1}{4m_W^4}(t^2+u^2-2s^2)-\frac{s}{m_W^2} \\
& +\frac{1}{2m_W^3}(t+2u)\Big(k.\Delta(p)+p.\Delta(k)+l.\Delta(r)+r.\Delta(l)\Big) \\
& +\frac{1}{2m_W^3}(u+2t)\Big(k.\Delta(r)+r.\Delta(k)+l.\Delta(p)+p.\Delta(l)\Big) \\
& +O(1)
\end{aligned}
\tag{J.27}
$$

Thus, from (J.21), (J.23) and (J.27) it is obvious that the choice (J.26) guarantees, beside an elimination of quartic divergences and also a cancellation of a part of quadratic divergences, namely of those corresponding to "dangerous" kinematical structures like $k.\Delta(p)$ etc. (these structures are potentially dangerous because they could not be compensated by means of diagrams involving a scalar exchange — cf. Eq. (5.72)). For the total contribution of the diagrams in Fig. 7, 19 and 20(c) (or, summarily, the graphs in Fig. 20) finally (using (J.25) and the relation $m_W^2/m_Z^2 = \cos^2 \vartheta_W$ - see (5.39)) we get

$$\begin{aligned}
\mathcal{M}_{7a} + \mathcal{M}_{7b} + \mathcal{M}_{19a} + \mathcal{M}_{19b} + \mathcal{M}_{20c} = & \\
= \quad \mathcal{M}_{20a} + \mathcal{M}_{20b} + \mathcal{M}_{20c} = & \\
= \quad -\frac{g^2}{4m_W^2}s + O(1) &
\end{aligned} \qquad (J.28)$$

The result (5.53) is thus proved.

Appendix K

LAGRANGIAN OF THE STANDARD MODEL

For the reader's convenience we summarize here the interaction Lagrangian of the standard model of electroweak interactions which we have deduced in Chapter 5 by means of a "diagrammatic method", i.e. by imposing the requirement of tree unitarity. The resulting interaction Lagrangian of the electroweak unification may be written as

$$
\begin{aligned}
\mathcal{L}_{int} &= \sum_f Q_f e \bar{f} \gamma^\mu f A_\mu + \mathcal{L}_{CC} + \mathcal{L}_{NC} \\
&- ig(W^0_\mu W^-_\nu \overleftrightarrow{\partial^\mu} W^{+\nu} + W^-_\mu W^+_\nu \overleftrightarrow{\partial^\mu} W^{0\nu} + W^+_\mu W^0_\nu \overleftrightarrow{\partial^\mu} W^{-\nu}) \\
&- g^2 \left[\frac{1}{2}(W^-.W^+)^2 - \frac{1}{2}(W^-)^2(W^+)^2 + (W^0)^2(W^-.W^+) - (W^-.W^0)(W^+.W^0) \right] \\
&+ g m_W W^-_\mu W^{+\mu} \eta + \frac{1}{2\cos\vartheta_W} g m_Z Z_\mu Z^\mu \eta \\
&+ \frac{1}{4} g^2 W^-_\mu W^{+\mu} \eta^2 + \frac{1}{8} \frac{g^2}{\cos^2\vartheta_W} Z_\mu Z^\mu \eta^2 \\
&- \sum_f \frac{1}{2} g \frac{m_f}{m_W} \bar{f} f \eta - \frac{1}{4} g \frac{m_\eta^2}{m_W} \eta^3 - \frac{1}{32} g^2 \frac{m_\eta^2}{m_W^2} \eta^4
\end{aligned}
$$

The term $\mathcal{L}_{CC}$ describes the interactions of weak charged currents and vector bosons $W^\pm$:

$$
\begin{aligned}
\mathcal{L}_{CC} &= \frac{g}{2\sqrt{2}} \sum_{l=e,\mu,\tau} \bar{\nu}_l \gamma^\lambda (1-\gamma_5) l W^+_\lambda + \\
&+ \frac{g}{2\sqrt{2}} (\bar{u}, \bar{c}, \bar{t}) \gamma^\lambda (1-\gamma_5) V_{CKM} \begin{pmatrix} d \\ s \\ b \end{pmatrix} W^+_\lambda + \text{h.c.}
\end{aligned}
$$

where V_{CKM} is the Cabibbo-Kobayashi-Maskawa unitary matrix (5.140). The term $\mathcal{L}_{NC}$ corresponds to the interaction of weak neutral currents and the vector boson Z:

$$
\mathcal{L}_{NC} = \frac{g}{\cos\vartheta_W} \sum_f (\varepsilon_L^{(f)} \bar{f}_L \gamma^\lambda f_L + \varepsilon_R^{(f)} \bar{f}_R \gamma^\lambda f_R) Z_\lambda
$$

where

$$\varepsilon_L^{(f)} = -\frac{1}{2} - Q_f \sin^2 \vartheta_W \quad \text{for} \quad f = e, \mu, \tau, d, s, b$$

$$\varepsilon_L^{(f)} = +\frac{1}{2} - Q_f \sin^2 \vartheta_W \quad \text{for} \quad f = \nu_e, \nu_\mu, \nu_\tau, u, c, t$$

$$\varepsilon_R^{(f)} = -Q_f \sin^2 \vartheta_W \quad \text{for an arbitrary } f$$

The neutral-current interaction may alternatively be written in the form

$$\mathcal{L}_{NC} = \frac{g}{\cos \vartheta_W} \sum_f \bar{f} \gamma^\lambda (v_f - a_f \gamma_5) f Z_\lambda$$

where

$$v_f = \frac{1}{2}(\varepsilon_L^{(f)} + \varepsilon_R^{(f)})$$

$$a_f = \frac{1}{2}(\varepsilon_L^{(f)} + \varepsilon_R^{(f)}),$$

that is

$$\begin{aligned} v_f &= -\tfrac{1}{4} - Q_f \sin^2 \vartheta_W \\ a_f &= -\tfrac{1}{4} \end{aligned} \Bigg\} \quad \text{for} \quad f = e, \mu, \tau, d, s, b$$

$$\begin{aligned} v_f &= +\tfrac{1}{4} - Q_f \sin^2 \vartheta_W \\ a_f &= +\tfrac{1}{4} \end{aligned} \Bigg\} \quad \text{for} \quad f = \nu_e, \nu_\mu, \nu_\tau, u, c, t$$

In the terms describing the self-interactions of vector bosons we have employed the notation

$$W_\mu^0 = \cos \vartheta_W Z_\mu + \sin \vartheta_W A_\mu$$

The following important relations are valid:

$$e = g \sin \vartheta_W, \qquad m_W / m_Z = \cos \vartheta_W.$$

References

[1] E. Fermi, Nuovo Cimento **11** (1934) 1; Z. Physik **88** (1934) 161.

[2] R. P. Feynman and M. Gell-Mann, Phys. Rev. **109** (1958) 193.

[3] N. Cabibbo, Phys. Rev. Lett. **10** (1963) 531.

[4] S. Weinberg, Rev. Mod. Phys. **52** (1980), 515;
     A. Salam, Rev. Mod. Phys. **52** (1980) 525;
     S. L. Glashow, Rev. Mod. Phys. **52** (1980) 539.

[5] S. L. Glashow, Nucl. Phys. **22** (1961) 579.

[6] A. Salam, in *Elementary Particle Physics* (Nobel Symp. No. 8), ed. N. Svartholm (Almquist and Wiksell, Stockholm, 1968) p. 367.

[7] S. Weinberg, Phys. Rev. Lett. **19** (1967) 1264.

[8] C. N. Yang and R. L. Mills, Phys. Rev. **96** (1954) 191.

[9] P. W. Higgs, Phys. Lett. **12** (1964) 132; Phys. Rev. Lett. **13** (1964) 508; Phys. Rev. **145** (1966) 1156;
     T. W. B. Kibble, Phys. Rev. **155** (1967) 1554.

[10] G. 't Hooft, Nucl. Phys. **B35** (1971) 167.

[11] J. M. Cornwall, D. N. Levin and G. Tiktopoulos, Phys. Rev. Lett. **30** (1973) 1268; (E) **31** (1973) 572.

[12] C. H. Llewellyn Smith, Phys. Lett. **46B** (1973) 233.

[13] J. M. Cornwall, D. N. Levin and G. Tiktopoulos, Phys. Rev. **D10** (1974) 1145; (E)**11**(1975) 972.

[14] S. D. Joglekar, Ann. Phys. (NY) **83** (1974) 427.

[15] E. S. Abers and B. W. Lee, Phys. Reports **9C** (1973) 1.

[16] J. D. Bjorken and S. D. Drell, *Relativistic Quantum Mechanics* (Mc Graw-Hill, New York, 1964).

[17] J. C. Taylor, *Gauge Theories of Weak Interactions* (Cambridge University Press, Cambridge, 1976).

[18] C. H. Llewellyn Smith, in *Phenomenology of Particles at High Energies,* Proc. XIVth Scottish Universities Summer School in Physics 1973 (Academic Press, New York, 1974) p.459.

[19] M. Jacob and G. Wick, Ann. Phys. (NY) **7** (1959) 404.

[20] V. B. Berestetskii, E. M. Lifshitz and L. P. Pitaevskii, *Quantum electrodynamics* (Nauka, Moscow, 1980) (in Russian).

[21] C. Itzykson and J.-B. Zuber, *Quantum Field Theory* (McGraw-Hill, New York, 1980).

[22] L. Durand and J. Lopez, Phys. Rev. **D40** (1989) 207.

[23] L. D. Landau and E. M. Lifshitz, *Quantum mechanics* (Nauka, Moscow, 1974) (in Russian).

[24] M. Gell-Mann, M. L. Goldberger, N. Kroll and F. E. Low, Phys. Rev. **179** (1969) 1518.

[25] Ta-Pei Cheng and Ling-Fong Li, *Gauge Theory of Elementary Particle Physics* (Clarendon Press, Oxford, 1984).

[26] D. G. Boulware, Ann. Phys. (NY) **56** (1970) 140.

[27] M. Veltman, Nucl. Phys. **B7** (1968) 637; **B21** (1970) 288.

[28] Review of Particle Properties, Phys. Rev. **D45**, Part 2 (June 1992).

[29] T. D. Lee and C. N. Yang, Phys. Rev. **128** (1962) 885.

[30] N. Cabibbo and R. Gatto, Phys. Rev. **124** (1961) 1577.

[31] Y. - S. Tsai and A. C. Hearn, Phys. Rev. **140** (1965) B721.

[32] H. Aronson, Phys. Rev. **186** (1969) 1434.

[33] K. J. Kim and Y. - S. Tsai, Phys. Rev. **D7** (1973) 3710.

[34] K. Hagiwara, R. D. Peccei, D. Zeppenfeld and K. Hikasa, Nucl. Phys. **B282** (1987) 253.

[35] G. 't Hooft, Nucl. Phys. **B33** (1971) 173.

[36] L. H. Ryder, *Quantum Field Theory* (Cambridge University Press, Cambridge, 1985).

[37] H. Umezawa, *Quantum Field Theory* (North-Holland, Amsterdam, 1956).

[38] S. J. Chang, *Introduction to Quantum Field Theory* (World Scientific, Singapore, 1990).

[39] R. Kleiss, in *Proc. 1989 Trieste Summer School in High Energy Physics and Cosmology.* The ICTP Series in Theoretical Physics, Vol. 6 (World Scientific, Singapore, 1990) p. 404.

[40] S. L. Adler, Phys. Rev. **177** (1969) 2426;
      J. S. Bell and R. Jackiw, Nuovo Cimento **A60** (1969) 47.

[41] H. Georgi and S. L. Glashow, Phys. Rev. Lett. **28** (1972) 1494.

[42] F. J. Hasert *et al.*, Phys. Lett. **46B** (1973) 121;
      C. Y. Prescott *et al.*, Phys. Lett. **77B** (1978) 347;
      J. E. Kim, P. Langacker, M. Levine and H. H. Williams, Rev. Mod. Phys. **53** (1981) 211;
      S. M. Bilenky and J. Hošek, Phys. Reports **90** (1982) 73;
      R. D. Peccei, *The Physics of the Standard Model*, preprint DESY 89-060 (1989).

[43] G. Arnison *et al.*, UA1 Collaboration, Phys. Lett. **126B** (1983) 398;
      P. Bagnaia *et al.*, UA2 Collaboration, Phys. Lett. **129B** (1983) 130;
      G. Arnison *et al.*, UA1 Collaboration, Phys. Lett. **129B** (1983) 273;
      P. Bagnaia *et al.*, UA2 Collaboration, Z. Phys. **C24** (1984) 1;
      C. Rubbia, Rev. Mod. Phys. **57** (1985) 699;
      P. Watkins, *Story of the W and Z* (Cambridge University Press, Cambridge, 1986).

[44] S. Weinberg, Phys. Rev. Lett. **27** (1971) 1688;
      J. S. Bell, Nucl. Phys. **B60** (1973) 427.

[45] C. Bouchiat, J. Iliopoulos and Ph. Meyer, Phys. Lett. **38B** (1972) 519.

[46] D. Gross and R. Jackiw, Phys. Rev. **D6** (1972) 477;
      C. P. Korthals Altes and M. Perrottet, Phys. Lett. **39 B** (1972) 546.

[47] R. Jackiw, in *Lectures on Current Algebra and its Applications*, eds. D. Gross *et al.* (Princeton University Press, Princeton, 1972).

[48] S. L. Adler, *Lectures on Elementary Particles and Quantum Field Theory* (eds. S. Deser *et al.*, MIT Press, Cambridge 1972).

[49] L. Rosenberg, Phys. Rev. **129** (1963) 2786.

[50] J. Hořejší, Czech. J. Phys. **B42** (1992) 241.

[51] J. Hořejší, Czech. J. Phys. **B42** (1992) 345.

[52] J. Hořejší, Czech. J. Phys. **B35** (1985) 820.

[53] A. D. Dolgov and V. I. Zakharov, Nucl. Phys. **B27** (1971) 525.

[54] J. Hořejší, J. Phys. **G12** (1986) L7.

[55] S. L. Glashow, J. Iliopoulos and L. Maiani, Phys. Rev. **D2** (1970) 1285.

[56] E. D. Commins and P. H. Bucksbaum, *Weak Interactions of Leptons and Quarks* (Cambridge University Press, Cambridge, 1983).

[57] M. K. Gaillard and B. W. Lee, Phys. Rev. **D10** (1974) 897;
      M. K. Gaillard, B. W. Lee and J. Rosner, Rev. Mod. Phys. **47** (1975) 277.

[58] Review of Particle Properties. Phys. Lett. **239B** (1990).

[59] J. D. Bjorken and S. L. Glashow, Phys. Lett. **11** (1964) 255.

[60] J. J. Aubert *et al.*, Phys. Rev. Lett. **33** (1974) 1404;
      J. E. Augustin *et al.*, Phys. Rev. Lett. **33** (1974) 1406;
      S. C. C. Ting, Rev. Mod. Phys. **49** (1977) 235;
      B. Richter, Rev. Mod. Phys. **49** (1977) 251.

[61] D. H. Perkins, *Introduction to High Energy Physics,* 3rd edition (Addison Wesley Publishing Co., New York, 1986).

[62] H. Georgi, *Weak Interactions and Modern Particle Theory* (Benjamin / Cummings, Menlo Park, California, 1984).

[63] M. L. Perl *et al.*, Phys. Rev. Lett. **35** (1975) 1489; Phys. Lett. **63B** (1976) 366.

[64] S. W. Herb *et al.*, Phys. Rev. Lett. **39** (1977) 252;
      L. M. Lederman, in *Proc. 19th Int. Conf. High Energy Phys.*, ed. G. Takeda (Phys. Society of Japan, Tokyo, 1978).

[65] F. Abe *et al.*, CDF Collaboration, FERMILAB-PUB-94/097-E.

[66] R. Fulton *et al.*, CLEO Collaboration, Phys. Rev. Lett. **64** (1990) 16;
      H. Albrecht *et al.*, ARGUS Collaboration, Phys. Lett. **234B** (1990) 409; Phys. Lett. **255B** (1991) 297.

[67] H. Albrecht *et al.*, ARGUS Collaboration, Phys. Lett. **229B** (1989) 175; Phys. Lett. **249B** (1990) 359;
      R. Fulton *et al.*, CLEO Collaboration, Phys. Rev. **D43** (1991) 651.

[68] Y. Nir, *The CKM Matrix and CP Violation*, preprint SLAC-PUB-5676, October 1991, and in *Elementary Particle Physics* (TASI-91) (World Scientific, Singapore, 1992).

[69] C. Albajar *et al.*, UA1 Collaboration, Phys. Lett. **262B** (1991) 163.

[70] M. Kobayashi and T. Maskawa, Prog. Theor. Phys. **49** (1973) 652.

[71] L. Maiani, Phys. Lett. **62B** (1976) 183;
L. Wolfenstein, Phys. Rev. Lett. **51** (1983) 1945;
L.- L. Chau and W. Y. Keung, Phys. Rev. Lett. **53** (1984) 1802;
H. Harari and M. Leurer, Phys. Lett. **181B** (1986) 123;
H. Fritzsch and J. Plankl, Phys. Rev. **D35** (1987) 1732.

[72] G. P. Lepage, *What is Renormalization?*, Cornell University preprint CLNS 89/970; in *Proc. TASI-89 Summer School*, Boulder, Colorado, 1989 (World Scientific, Singapore, 1990).

[73] M. S. Chanowitz, Ann. Rev. Nucl. Part. Sci. **38** (1988) 323.

[74] J. F. Gunion, H. E. Haber, G. Kane and S. Dawson, *The Higgs Hunter's Guide* (Addison-Wesley Publishing Co., New York, 1990).

[75] W. A. Bardeen, C. T. Hill and M. Lindner, Phys. Rev. **D41** (1990) 1647;
R. S. Chivukula, A. G. Cohen and K. Lane, Nucl. Phys. **B343** (1990) 554;
A. Dobado and M. Herrero, Phys. Lett. **228B** (1989) 495;
J. F. Donoghue and C. Ramirez, Phys. Lett. **234B** (1990) 361;
S. Dawson and G. Valencia, Nucl. Phys. **B352** (1991) 27.

[76] F. Halzen and A. D. Martin, *Quarks and Leptons: An Introductory Course in Modern Particle Physics* (John Wiley & Sons, New York, 1984).

[77] G. Kane, *Modern Elementary Particle Physics* (Addison-Wesley Publishing Co., New York, 1987).

SUBJECT INDEX

www.ingramcontent.com/pod-product-compliance
Lightning Source LLC
Chambersburg PA
CBHW050642190326
41458CB00008B/2382

* 9 7 8 9 8 1 0 2 1 8 5 7 7 *